Y0-BVO-595

Temperate Conquests

Temperate Conquests

Spenser and the Spanish New World

David Read

WAYNE STATE UNIVERSITY PRESS
DETROIT

Manufactured in the United States of America.
04 03 02 01 00 5 4 3 2 1

Library of Congress Cataloging-in-Publication Data
Read, David, 1956–
Temperate conquests : Spenser and the Spanish New World / David Read.
p. cm.
Includes bibliographical references and index.
ISBN 0-8143-2872-5 (alk. paper)
1. Spenser, Edmund, 1552?–1599. Faerie queene. 2. Didactic poetry, English—History and criticism. 3. Spenser, Edmund, 1552?–1599—Knowledge—America. 4. Spain—Colonies—America—History—16th century. 5. Epic poetry, English—History and criticism. 6. America—Discovery and exploration—Spanish. 7. English poetry—American influences. 8. English poetry—Spanish influences. 9. Colonies in literature. 10. America—In literature. I. Title.
PR2358.R38 2000
821'.3—dc21 99-33276

An earlier version of chapter 4 appeared as "Hunger of Gold: Guyon, Mammon's Cave, and the New World Treasure." *English Literary Renaissance* 20.2 (1990): 209–32. Reprinted by permission of the editors.

IN MEMORY
OF MY FATHER

Contents

Acknowledgments

This book seems to have traveled almost as long as any of Spenser's wandering knights, and encountered nearly as many perils (some of which might even be amenable to allegorical reading). Rather than trace that journey in detail, I want to acknowledge those who have offered support and solace along the way, and who may be surprised to hear that they have had any effect on the work that follows. But what I am honoring here is what I value most about them, their friendship. First, I want to thank Janel Mueller, who kept me in mind when there was no need for her to do so and who helped me in ways large and small to keep going as a scholar. Without ever having taken a class from her, I still account myself her student. Then there are people who I associate with particular institutions, though some of them have since moved on: at the University of Chicago, Michael Murrin, the late John M. Wallace, Janice Knight, Josh Scodel, and David Bevington (for at least one timely good office); at the College of St. Francis, Randy Chilton; at Marquette University, Tim Machan; and at the University of Missouri, Don Anderson, Martin and Sandy Camargo, Bill Dawson, Howard Hinkel, Haskell Hinnant, Doug Hunt, Bill Kerwin, Tim Materer, Catherine Parke, Tom Quirk, Trish Roberts-Miller, Tom Stroik, Nancy West, and Gilbert Youmans. It has also been my pleasure at Missouri to come in contact with the professionalism and exemplary Renaissance scholarship of Charles Nauert, and to share the nuances of Italian-to-English translation with Glenn Pierce, Alan Thiher, and Mary Jo Muratore. I was glad to have encountered, and will continue to miss, the enthusiasm and kindness of the late Paul Casey. I think of others with whom I have passed rewarding time in one ivory tower or another: Shelly Smith and Jeremy Drelich, the late Randy Chauvin, George

Klawitter, Phil Herring, Tom and Pam Culviner, Jeff and Cathy Nelson, Jim and Sylvia Hecimovich, Greg and Sheila Meyerson, Tina Reithmayer, and Charlie Ross.

This is a different and better book because of the extraordinarily meticulous attention it was given by the readers at Wayne State University Press. My appreciation goes as well to *English Literary Renaissance* and its editor, Arthur F. Kinney, for giving me permission to reprint most of an essay that first appeared in that journal in chapter 4 of this book.

Finally, I acknowledge with love and immense respect my late father, Dr. Edward J. Read, who made his exit from this world shortly before this book could enter it. "But were it not, that *Time* their troubler is, / All that in this delightfull Gardin growes, / Should happie be, and haue immortall blis." I am thankful to my mother, Marian, and my siblings Becky, Paul, and Joel, for performing so well and so readily the often difficult task of being "family." My deepest gratitude, though, goes to my wife, Sarah, and daughters, Anna and Molly, for what they have always given unstintingly and unconditionally, "contrayr to *Mutabilitie*."

Introduction: Spenser's Horizon

This book is about Edmund Spenser's response as a writer—especially as the writer of *The Faerie Queene*—to the colonial movements that were so much a part of the European scene during his career, as well as to the aspirations of the English to be among the prime movers in the sixteenth-century project of overseas expansion. Within the poem Spenser offers his own considered contribution to what is in effect a "national dialogue" on the character and progress of English colonialism during the Elizabethan period. My argument thus traverses some familiar terrain, but perhaps it does so along an increasingly neglected path. For much of the recent criticism on Spenser has centered less on *The Faerie Queene* than on *A View of the Present State of Ireland,* a work that remained unpublished until thirty-four years after his death and until only a few years ago was treated as one of the less significant items in his corpus.[1] *A View* has attracted this attention because it seems such a fortuitous exposition of one of the perennially vexing problems in the study of Spenser: his ambiguous status as both an Elizabethan artist of great complexity and an apparently unreserved apologist for Elizabethan colonial policy. "Faeryland is not merely an ideal Elizabethan England," according to Willy Maley, "but a site of struggle for Tudor imperial aspirations."[2] Richard McCabe has incisively described the great paradox that shapes both Fairyland and Spenser's entire career: "Although commonly remembered as *the* Elizabethan poet *par excellence,* he spent remarkably little time at the court of 'Gloriana'. All genteel pretensions to the contrary, he was not a courtier but a colonist. The years of his poetic and political maturity were passed not at the centre but on the periphery of the Elizabethan state, not in the city but in the wilderness."[3] This problem intersects with the intense interest among literary critics and cultural historians in all things colonial, which reached its apogee in 1992 during the Columbian quincentennial and which continues to run a lengthy orbit.

The specific upsurge of interest in Spenser's colonialism, though, almost certainly dates back to Stephen Greenblatt's famous discussion of the Bower of Bliss in *Renaissance Self-Fashioning*,[4] which has largely determined—if not frozen in place—the contemporary interpretation of Spenser's relation to English colonialism.[5] While Greenblatt's general critique of the colonial-in-the-literary and the literary-in-the-colonial influenced or at least opened the way for a great deal of recent criticism,[6] his interpretation of *The Faerie Queene* as a display of "the passionate worship of imperialism" and of Spenser as "our originating and preeminent poet of empire"[7] had an oddly dampening effect on subsequent efforts to examine Spenser in a similar light.[8] During the height of the New Historicism in the late 1980s and early 1990s, there was nothing in Spenser studies remotely like the explosion of articles about *The Tempest* as a colonial text, which Greenblatt also helped to detonate.[9] The effect of Greenblatt's argument was to create a generalized version of "imperial" Spenser functioning almost independently of his poetic output; citations of one or two significant episodes from *The Faerie Queene* were enough to confirm Spenser's place as the great apologist for an Elizabethan dream of transoceanic monarchy.[10]

Though their debt to the "school of Greenblatt" is often patent, recent readers of *A View* have also done a great service to Spenser studies in bringing back some energy and concrete complexity to what had become a nearly static conception of Spenser, one that postcolonial studies had inherited without much further scrutiny.[11] These readers are not interested so much in exculpating Spenser from the charge of colonialist complicity as in recovering a cohesive sense of the intricacies of his relationship to the Elizabethan social order; thus Andrew Hadfield's statement that

> [Spenser's] work is defined by the Tudors' attempt to expand their boundaries and unify a nebulously conceived ideal of Britain, as well as exploit and subdue other nations and cultures. At the same time, Spenser's works participate in and reflect upon that enterprise in an active way, as Spenser himself participated in English colonial expansion via his career as a government official in Ireland.[12]

This is typical of the latest commentary on Spenser's Irish connections.[13] His writing is treated as inseparable from a peculiar and peculiarly troubled juncture in the network of Elizabethan history, culture, and policy—a juncture whose details will always be under debate, since the topic still manages to carry a violent charge for anyone concerned with Anglo-Irish conflict over the centuries. Spenser will always have a role in "the troubles."

As far as that debate goes, I have little argument with the approach of Hadfield and his peers, but I hope this book will serve, by way of addition and amendment, to return students of Spenser to a broader context for thinking about his "active" involvement with English colonial activity in a way that does not simply lead back to the abstraction of Spenser as the "originating and preeminent poet of empire." The looming peril at present is that *The Faerie Queene* will be superseded by *A View* as the primary historical document of Spenser's political and ethical vision. Analysis of the poem in the recent Irish studies leans heavily toward the three books added in the 1596 edition and especially to Book 5, with its obvious allusions to the persons and events of the English campaign in Ireland during the 1580s—the point being, of course, to find the most cogent links between *The Faerie Queene* and *A View*. Fair enough, yet I will argue that Spenser's most dedicated thinking about the general problem of English colonialism (a problem suitably articulated in the first sentence of the quotation from Hadfield above) actually occurs in the 1590 edition and that this thinking mainly occurs in Book 2, which seems to have replaced Book 5 as the most underrepresented section of the poem in current criticism. Moreover, the locus of the "colonial problem" in Book 2 is not Ireland but the Spanish New World. I will extend this argument quite a bit further than it has been taken by previous critics of Book 2, including those associated with the New Historicism. My thesis entails several more claims: Spenser was familiar with the accounts of the Spanish conquest of the Americas that had been disseminated in England beginning in the 1550s; the issues he found in those accounts deeply engaged his poetic imagination; and he sought in a sustained way to allegorize those issues in relation to the virtue that he found most appropriate to them—namely, the virtue of temperance.

This is not to discount the importance of Ireland in shaping Spenser's artistic purposes (though I will have something to say in chapter 5 about critics' overzealousness in finding allusions to Ireland in *The Faerie Queene*). However, Spenser thought and wrote about the problems faced by the "Englishman abroad" in cosmopolitan, indeed, in international, terms; Ireland did not constitute the outer boundary of his capacity for imagining those problems. It is a truism that Ireland provided a field on which the English could test policies and practices that they would pursue on a much larger scale in their later colonial ventures. These policies and practices are usually understood as being directed toward the natives, but the members of Spenser's generation would have understood them as being equally directed toward a dangerous foreign power that had its own reasons for gaining a foothold in Ireland.

My path here leads, for a moment, close to an infamous parcel of Irish territory: Smerwick in County Kerry on Munster's far western coastline, a place that figures both physically and morally in Spenser's landscape at a formative stage in his career, not long after the publication of *The Shepheardes Calendar* but before he had taken up permanent residence in Ireland. It also represents a perennial crux in the effort to fill in the details of Spenser's biography. Here was the site of the Fort del Oro, which in November 1580 was garrisoned with a poorly organized expeditionary force of Italian and Spanish mercenaries sent under the auspices of Pope Gregory XIII to assist the Irish during the Desmond rebellion. Arthur Lord Grey de Wilton, Elizabeth's recent appointee as Lord Deputy of Ireland and Spenser's employer at the time, successfully laid siege to the fort; having secured the unconditional surrender of the garrison, he ordered almost everyone in it—between four hundred and seven hundred people, including Irish camp-followers—to be put to death.[14] Sir Walter Ralegh, Spenser's staunchest supporter later in the 1580s, was directly involved in the massacre. Whether or not Spenser himself was physically present, his position as Lord Grey's private secretary suggests that he could not have been very far from the scene. He certainly was aware of what happened there.

Grey's stint as Lord Deputy and, accordingly, Spenser's career with him lasted only two years; the action at Smerwick was an egregious instance of Grey's obdurate and frequently counter-productive militancy in dealing with Irish unrest, and early evidence that his government would never achieve any degree of effectiveness apart from its brutal suppression of rebellion. This did not prevent Spenser from excusing Grey's conduct in *A View*,[15] and evidently in Book 5 cantos 11–12 of *The Faerie Queene* as well. Though the soldiers in the papal force seem to have been mainly Italian in origin, *A View* has Eudoxus, the "English" participant in the dialogue, refer to the massacre as "that sharp execution of the Spaniards."[16] For Spenser, the members of the doomed garrison were at least symbolically Spanish, if not actually so. They were conquistadors, harbingers of King Philip's grand design to bring England forcibly within Spain's empire, numbering nearly as many as the soldiers in the army that Cortés had marched across Mexico sixty years before, gathering up native rebels to support his cause. (In fact several English-language accounts of Cortés's Mexican campaign were available in print prior to 1580.) From Spenser's perspective, Grey's actions at Smerwick were justified as being, so to speak, "the best defense" against an obvious threat to the kingdom, a threat in which Ireland figured less as a point of arrival than as one of departure.[17]

From Smerwick, the horizon broadens: the problem of Spanish imperialism becomes, literally, as wide as the Atlantic, and the signs (both actual and imagined) of Spanish designs toward hegemony in Europe and the Americas become pervasive, culminating, of course, in the Armada of 1588. I claim that for Spenser, an event like Smerwick and one like the Armada formed part of a single intellectual fabric in which it was plausible—even necessary—to take Spain's activities in the New World as part of the weave. In the following chapters I make the case that Spenser wove an informed knowledge of the Spanish New World into the fabric of the first three books of *The Faerie Queene*. While that case is inevitably circumstantial, it can be made with some force: apart from the textual evidence I explore, one can look to Spenser's continuing association with individuals involved in English colonial projects in the New World—not only Ralegh, but also Sir Philip Sidney, the poet Edward Dyer, and the Earl of Leicester.[18] Indeed, it would be surprising if a poet so often concerned with the *outward* movements of his implicitly English heroes from the court of the Fairy Queen did not make some imperialist gestures. However, such gestures—those of Spenser's associates as well as his own—are always made under the long shadow cast by Spain's empire, an empire firmly established, still expanding, and always threatening.

Paradoxically, this empire, so rich in land, gold, and subjects, was almost unavoidably admirable as well. Spain's history of success in the New World was paradigmatic in that the other European polities were reduced merely to imitating what had already been done better and on a broader scale. Charles Nicholl has observed in his fascinating book on Ralegh's first voyage to Guiana that "one of the curiosities of the new English interest in America . . . is that it entailed a curious rapprochement with the idea of 'Spanishness'—a sense that the Spaniards were the only role models for American exploration, and that Englishmen bound for America must emulate them, must in a sense *become* Spanish."[19] My own sense of this rapprochement is that it is not particularly straightforward, but involves strategies of mediation and deflection that surface regularly enough in Elizabethan texts to be considered characteristic. A more precise formulation might be this: the English emulate Spanishness by denaturing it, by making it become English. The peculiar problems associated with this process take up many pages in the following chapters.

It is one thing, though, for a writer to think about these issues in political terms, on the scale of "current events" near to or far from one's personal concerns, and another to interpret them artistically. This problem is compounded in a poem as self-consciously and multifariously an "objet d'art" as is *The Faerie Queene*. The task for the poem's critics is to understand Spenser's political interests within the dimension of his poetics—the dimension that, after all, remains primary. In his own pursuit of this task, Gordon Teskey has nicely described the "entire project" of the poem as "an improvisatory quest for the formal patterns of relationship which sustain judgments of value in human affairs."[20] The emphasis here falls on "formal," which for Teskey is primarily a matter of visual forms. The poem's project thus implies a distinct tension in regard to purpose, which Teskey posits as follows:

> first . . . [Spenser] was sensitive to the nuances of social and political life in a more immediate way than was the case even with Dante or Milton; and second . . . he was not primarily a narrative poet but a poet whose main concern was to think. He

> thought in subtle, allusive, indirect, and intuitive ways about problems too complex to be dealt with by entirely rational means, problems we might describe as demanding an associative rather than an algorithmic approach.[21]

This passage provides a useful characterization of the basic problem of Spenser's methodology in *The Faerie Queene:* namely, its constant "associative" shifting between the abstract and the concrete, the aesthetic and the pragmatic, which has often defeated the efforts of Spenser's interpreters to categorize the poem as being about either metaphysics or "physics" (by which I mean "sublunary things," in the Elizabethan sense).

A way around this problem is to follow Spenser's cue in the letter to Ralegh and understand the poem as fundamentally a meditation on conduct, as such conduct is summed up in "virtuous and gentle discipline."[22] The topic of conduct is a familiar feature of both popular and intellectual discourse in the sixteenth century; it is also deeply amenable to allegory because it presents itself under so many aspects: outward, inward, public, private, political, spiritual, collective, individual, past, present (and perhaps even future). For Spenser, the topic has the advantage of allowing the poet to speak relevantly to his contemporaries about serious moral and ethical issues without needing to rely on fixed historical referents in order either to tell a story or, in Teskey's terms, "to think." Such referents are never absent from *The Faerie Queene,* but Spenser as a historical allegorist aims for what might be termed "pertinent generality," stressing *how* over *what* in his querying of human affairs. As Michael O'Connell remarks,

> Spenser more often refers to the contemporary world allusively, through momentary indications of a moral relationship between the poem and its political context. At certain points, particularly in the second half of *The Faerie Queene,* this concern for contemporary events and issues does grow into full-scale allegorization. But it is more accurate to speak of the historical dimension of Spenser's poetry, a term that includes the full range of allusion, satire, symbolic characterization, historical catalogue, and topical allegory.[23]

This broadly construed "historical dimension" occupies a middle ground between the roman à clef and the realm of eternal forms, ground on which an artist like Spenser can operate productively and, so to speak, out of danger. The reader need not connect Redcross to Leicester or Guyon to Ralegh or Artegall to Lord Grey (or, for that matter, Gloriana and Belphoebe to Elizabeth) to understand that the actions of these characters point to questions of consequence for English people in the sixteenth century: How should the English "gentleman" behave in matters of religion, politics, sexuality, or in the many forms of civic and domestic life? How should that "gentleman" behave when faced with new or alien forms of life? How should a whole kingdom of such "gentlemen" behave?

These are guiding questions in *The Faerie Queene,* which are susceptible to both concrete and abstract responses, but in either case responses that carry considerable weight when Britain itself is, in David Baker's words, "less a fixed and distinct domain than an ontological predicament."[24] In Spenser's poetic universe, conduct and social identity are mutually defining: to behave in a certain way is to be English; to be English is to behave in a certain way. Problems arise when one or both of the terms in the relationship is ill-defined, or not defined well enough to be prescriptive. As the historian Keith Robbins has noted,

> A country's conception of itself is never static. The "space" it believes itself to occupy shifts constantly. Institutions and ideas deemed to be characteristic only become so by comparison and contrast with what is believed to prevail elsewhere. In turn, that external world reinforces or destroys indigenous assumptions and aspirations. What is "core" and what is "periphery" is likewise fragile. Perceptions of order and location, deemed immutable, quake and disintegrate in the light of new knowledge and fresh contexts.[25]

Certainly one of the more mutable conceptions within England in the late sixteenth century involved the conduct of the English colonist. As murky and unstable as the Irish situation was during Spenser's residence there, it could at least be considered a real situation, involving actual colonists taking over verifiable territory. But when Spenser published the first three books of *The Faerie Queene* in 1590, Ralegh's American colony was no more tangible than Fairyland; for all practical purposes, Virginia might as well be an allegorical representation of the state of English overseas empire.[26] At the same time—and as I hope the next chapter makes clear—the impulse to treat colonization, and especially colonization across the Atlantic, as one of the defining activities of the "gentleman" was very strong in the literary and political culture of Spenser's day. In making a place for colonial matters within his larger allegorical design, Spenser faces the difficulty of depicting a "truth" that has yet to be revealed. The prevailing model for expansionism belongs to the archenemy of the kingdom, and the Elizabethans have no analogous model to offer in its place. Lacking positive alternatives, Spenser faces the puzzle of portraying English colonial conduct primarily by what that conduct is not.

Given this puzzle, it is perhaps not surprising that the most sustained representation of colonial activity in *The Faerie Queene* occurs in Spenser's book "Of Temperaunce." For the virtue of temperance is characterized by abnegation—by *not* doing certain things, acting in certain ways, or following particular paths. Commenting on Book 2, René Graziani says that its references to the New World "accurately signpost the direction of the book: nature, or more specifically human passions, and their need to be controlled."[27] Likewise, Thomas H. Cain argues that Book 2 "warns Elizabethans to avoid Spanish excesses in the Americas."[28] Self-control, avoidance, hewing to the mean—these are the leading aspects of the virtue of temperance. At the same time, they are also pieces of the puzzle mentioned above. The English colonist as Spenser imagines him is essentially the temperate gentleman, celebrated for what he chooses never to do. The challenge for Spenser is to transform this *lack* of activity, which might also be extrapolated into a lack of identity, into an authentic virtue that is fundamental to the teaching of "gentle discipline."

What exacerbates this challenge is that, as previously suggested, there is no English empire for this "poet of empire" to celebrate. The only empire that Spenser and most of his contemporaries knew about in detail, other than the old Roman one, was the one that gave rise to *la leyenda negra* and the Armada. Spenser is unable to allegorize the character and activity of the English colonist from a convenient position, since the "foundation" of Elizabethan overseas policy at the time of *The Faerie Queene*'s composition is a loose collection of manuscripts, pamphlets, quarto volumes, and occasional verse, promoting deeds that mostly have yet to be done and reflecting (often wistfully) on the past accomplishments of a hostile foreign power.[29] Lost in the recent "colonialist" account of Spenser is a sense of a particular poet struggling with a practical literary problem: how to give form and meaning to—how to find instruction and edification in—such an unstable mix of rhetoric and reality.

This book, then, is an effort to retrieve Spenser's engagement with colonialism in *The Faerie Queene* at the level of poetic composition. The challenge in this approach is to balance contextual investigation with close reading in a way that acknowledges the importance of both modes to interpreting Spenser's work.[30] I hope to capture what O'Connell in the passage cited earlier calls *The Faerie Queene*'s "historical dimension," while avoiding the reductive understanding of historical allegory that has hampered some of the earlier attempts to read the New World into the poem.[31] The main flaw in these attempts lies with the assumption that Spenser's historical allegory is primarily concerned with specific events and persons, so that in effect a single "key" will unlock the meaning of a given passage in the poem. Instead, I believe one does more justice to this allegory by understanding that it is concerned with problems raised within a historically situated discourse for which Spenser provides a receptive audience.

The text will show how Spenser struggles to digest the discourse of Elizabethan colonialism not into an overt (or, for that matter, covert) political program, but into a sustained, meaningful poetic narrative that is both "plausible and pleasing," as the letter to Ralegh puts it.[32] The discussion will turn on a detailed analysis of Book 2 as the main place where this activity occurs, though my analysis entails considerable lateral movement, since I am also concerned with the ways that Spenser has carried the elements of a specific intellectual milieu into his poem. To that end, chapter 1 offers an account of the "study" of New World colonialism in England in the 1570s and 1580s and identifies the probable sources and conduits of Spenser's interest in the subject. The next five chapters work their way through the major narrative moments in Book 2, with occasional digressions into other books of *The Faerie Queene.* Chapter 2 deals with the ideological rationale for Guyon's quest, as it is signaled in the Ruddymane episode and in Guyon's reading of the Elfin chronicles at the House of Alma. The next chapter points out the ways in which the landscape of Book 2 functions as an allegorical representation of a colonial "geography"; it then analyzes Guyon as a type of anti-conquistador whose activities are inevitably in contrast with those of the various knights who populate that landscape, most of whom reflect contemporary cultural stereotypes of the conquistador. Chapter 4 develops this notion in relation to the first of Guyon's great ordeals, the Cave of Mammon, where Guyon confronts many obstacles associated with the Spanish quest for treasure in the New World. Chapter 5 addresses the Maleger episode in canto 11, showing that "Irishness" can be imposed onto the poem in a way that does not sufficiently account for Spenser's perspective on colonialism. The following chapter takes up the ultimate crux in Book 2, the Bower of Bliss, to argue that the Bower's destruction is both less momentous and more ambiguous than it is made out to be in Greenblatt's influential account of the episode. The book concludes with a reflection on the difficulty of determining the influence of Spenser's poetry—and of Elizabethan poetry generally—within large-scale historical and cultural processes, suggesting finally that a work like *The Faerie Queene* is by its nature divorced from actual politics and that its political impact, as such, is restricted to the field of literature. From this perspective, Spenser can be seen as registering rather than arbitrating the question of Elizabethan empire, and Book 2 as offering a testament to the conceptual difficulties surrounding English colonialism rather than providing a definitive vision of England as an imperial nation.

I do want to emphasize those "conceptual difficulties"; as with so much else in *The Faerie Queene,* there is nothing easy about Spenser's allegorization of the colonizing Englishman. Spenser's vision never becomes definitive because he has to confront from within that vision a range of political, ethical, and artistic obstacles that compromise any notion of empire that is supposed to emerge from the poem. Richard Helgerson, examining the conflict in Spenser's poetry between impulses toward both classical epic and chivalric romance, calls *The Faerie Queene* "a poem divided against itself."[33] Helgerson conceives of this division in terms of a broad dualism, but it might be more accurate to say that the poem is multiply divided and that its quest for meaning is always moving toward complex rather than simple oppositions.[34] Book 2 can also be understood as divided in the sense that it reflects Spenser's struggle to articulate, within his own intricate aesthetic, an issue that pressed with some urgency on him and his contemporaries; namely, England's claim to have the requisite power (and virtue) to *be* a colonizing nation. Book 2 also reflects the historical struggle of the Elizabethans to find their own position and prerogative among the competing claims of a larger Europe. Neither of these struggles lends itself to unproblematic conclusions any more than does the Elizabethans' ill-starred effort to plant the English nation in the New World.

As my working text of *The Faerie Queene,* I have used A. C. Hamilton's edition in the Longman Annotated English Poets Series (London: Longman, 1977), which reprints the text of J. C. Smith's 1909 Oxford edition. In quoting from sixteenth-century sources in this book, I have parted from my practice in earlier work and have modernized spelling in every text except *The Faerie Queene.* Both the printed titles and the original punctuation of the sources have been left intact, and I have not altered Elizabethan vocabulary where respelling would effectively substitute a different word (e.g., "sith" to "since"). But I have decided that consistency and readability outweigh the need for reproducing Elizabethan orthography exactly and hope that the reader will feel well served by this decision.

CHAPTER 1

The Elizabethan Projection

As, for example, I may speak (though I am here) of Peru,
and in speech digress from that to the description of Calicut.

Sir Philip Sidney, *A Defence of Poetry*

Who fares on sea, may not command his way,
Ne wind and weather at his pleasure call:
The sea is wide, and easie for to stray;
The wind vnstable, and doth neuer stay.

The Faerie Queene 2.6.23.2–5

I am not, let me insist, concerned with events, but with
ideas, for history is often more instructive as it considers
what men conceived they were doing rather than what, in
brute fact, they did.

Perry Miller, *Errand into the Wilderness*

In the proem to Book 2 of *The Faerie Queene,* Spenser claims to "vouch antiquities, which no body can know" (Pr.1.9), to describe a world—Fairyland—for which there is no prior evidence, however fragmentary. He offers "certaine signes" (Pr.4.2) of a place that has no past history by which to verify itself. What allows Spenser to make this claim with such confidence and to avoid the judgment that his work is only "th'aboundance of an idle braine" (Pr.1.3)? The proem's second stanza, as has often been noted, offers three possible answers: Peru, the Amazon, and Virginia. At the moment of their discovery, all three were (from the European perspective, of course) ahistorical entities, belonging to the present and, most crucially, to the future of their discoverers. Moreover, the astonishing reality of these new regions and the superlatives they engendered made questions of "history" and "antiquity" seem, for the moment, insignificant. Peru was overwhelmingly *there;* how could anyone ever question its existence?

Thus it is with the "antique history" of Book 2. Spenser is not writing the past history of regions that are known to the visionary poet in the present and will be revealed to those less clairvoyant in the future. He is writing the history of the future itself, of the as-yet-undiscovered inheritance of England, and he answers his skeptics with reasoning that might be paraphrased this way: "Peru exists, though no one in Europe suspected its existence; with it rests the prosperity of Spain now and in time to come. If Peru, why then not a greater, richer, more marvellous land, at present unknown? Why not a prosperous future there for the English nation?"[1] This would appear to be the substance of Spenser's message to the Queen in stanza 4: "And thou, O fairest Princesse vnder sky, / In this faire mirrhour maist behold thy face, / And thine owne realmes in lond of Faery, / And in this antique Image thy great auncestry" (Pr.4.6–9). The "antique Image" presents not only an "auncestry" (as in canto 10 of Book 2) but also a posterity to the beholder: Fairyland is there; rich with promise, waiting to be possessed, and hovering on the far side of the "faire mirrhour" into which Elizabeth gazes at the present expectant moment.[2] The English people of this new and wiser age have been granted an oracular guide—Spenser himself—to hold up a looking glass that is also a window into the future.[3] The proem to Book 2 introduces not so much a discovery or rediscovery as a pre-discovery. Though Spenser may seemingly assume the mantle of prophet here,[4] his indication of a genuinely New World already bestowing its abundant gifts on the Old World lends his words the quality of *prophetia ex eventu;* the success of the Elizabethan adventure is (or ought to be) a foregone conclusion.

Spenser's stance on these places in England's future is, of course, a matter of book learning, which is to say, something close to "experience" in the sixteenth-century context. Most Europeans outside of the Spanish sphere of influence—most Spaniards, in fact—were confined to viewing the New World as something that existed within the textual record that developed in the wake of the Conquista. In an environment where writing was compelled by the limitations of concrete human experience to become a substitute for that same experience, there existed a readiness to confuse writing (and reading) with doing, with actions undertaken on a different physical plane. This confusion gave rise to a belief that writing possessed what might be called "historical presence"; in other words, that a purportedly descriptive text not only described but also contained men's actions.

Thus, it was possible for chroniclers who never set foot outside of Europe—Peter Martyr and Francisco López de Gómara are prominent examples—to appear to occupy significant positions on the same historical field as Columbus or Cortés, functioning as direct observers of events occurring within their own textual domains rather than as mere recorders of external facts. It was also possible for a pretender like Vespucci to pass off his fabricated letters as truthful accounts of his exploits and to see his claims of having "found a continent" accredited by the European intelligentsia.[5] Yet the cultural assumptions that allowed Vespucci's impostures to meet with acceptance could also provide an impetus to someone like the great "apostle of the Indies," Bartolomé de Las Casas, whose writings are among the few that accurately can be said to have altered Spanish imperial policy. The debate over the truthfulness of his *Brevissima Relación de la destruycion de las Indias* has continued to the present day, but whether readers regard this tract as a factual document or simply as inflammatory propaganda, they remain vividly aware of Las Casas "acting" through the text to improve the material conditions of the native populations of Central America and the Caribbean.

There is little practical difference between Las Casas's intentions in writing the *Brevissima Relación* and in serving for so many years as a missionary to the New World; the salvation of the natives was always his objective, single-mindedly pursued. But there is a significant difference in genre, so to speak. The reality of Las Casas's work in the Indies was shared with an audience not much larger than the recipients of his good offices, whereas the "reality" of his pamphlets depended crucially on the confirming and valorizing response of his readers in the Old World. Las Casas himself was keenly aware of this distinction; in a remarkable passage from his *Historia de las Indias,* he placed the onus for atrocities against the Mexicans on what may seem like an unlikely source:

> You see how Cortés has the world deceived and those who read his false history are to blame for not reflecting that Indians had not offended us, that Indians are not Moors or Turks who plague and mistreat us; in short, they [the readers] are to blame for accepting the written word: Cortés killed and Cortés won, he conquered—as they say—many nations, he plundered and stacked gold in Spain and became the Marquis del Valle. Readers are to blame for this, especially the learned among them.[6]

To cast such blame on "accepting the written word" is to privilege that word and its acceptance in a very heightened way and to elevate the history of a text's reception to a drama of the most serious order. According to Las Casas, the text, its writer, and its readers mutually share the power to shape history. Whether any text could bear the weight of this claim is highly doubtful; at the same time, the claim renders moot the question of whether or not the events described in a text represent the objective truth. It is sufficient if the text's readers accept them as such, since readers are also the "makers" of truth. The field of history and of history-making no longer belongs solely to the historians and the conquistadors; it becomes the province of philosophers, scholars, priests, and poets, the last of which would find themselves in a position to prove for good or ill Sidney's claim in *A Defence of Poetry* that they are naturally superior to historians.

"Projection" is an apt term for the phenomenon I have been describing. This word was current in Elizabethan England for the casting of plans or schemes, but it had also already acquired its technical meaning in cartography as the laying out of the earth's sphere onto a flat surface—to wit, the "Mercator projection."[7] A particularly complex use of the term occurs in Fulke Greville's *Life of Sir Philip Sidney,* where Greville describes Sidney's abortive attempt to become a colonist in North America: "Nevertheless as the Limbs of *Venus* picture, how perfectly soever began, and left by *Apelles,* yet after his death proved impossible to finish: so that *Heroical* design of invading, and possessing *America,* how exactly soever projected, and digested in every minute by Sir *Philip,* did yet prove impossible to be well acted by any other mans spirit than his own."[8] Here the design that has been "projected" suggests not only a plan of action but also the drafting or visualization of a work of art. Colonization in this account becomes an aesthetic endeavor in which artists naturally perform better than laymen. For Greville, at least, America is as much a medium of creative expression as a place.

In many of the sixteenth-century works that attempt to interpret the content and meaning—the signs—of the New World, there is this metaphorical projection of the surface of the earth onto the surface of the text. Each text is, in this sense, a map, but is also the surface onto which individuals project their own wills and desires, under the assumption that they will go on to "make" history. One is reminded of Merlin's looking glass in Book 3 of the poem, in which Britomart sees the image of her future husband Artegall:

> It vertue had, to shew in perfect sight,
> What euer thing was in the world contaynd,
> Betwixt the lowest earth and heauens hight,
> So that it to the looker appertaynd;
> What euer foe had wrought, or frend had faynd,
> Therein discouered was, ne ought mote pas,
> Ne ought in secret from the same remaynd;
> For thy it round and hollow shaped was,
> Like to the world it selfe, and seem'd a world of glas.
> (3.2.19)

The implicit hope of the "looker," a hope far more serious than mere wishful thinking, is that the projection will go beyond representing the world and become "the world it selfe"—that description will be transformed into action. Whether or not this hope can ever bear fruit, the process of projecting goes on notwithstanding, contributing in turn to a larger process of cultural definition in which societies (both consciously and unconsciously) make sense of what they are and what they aspire to be, devising around themselves a world to accommodate the definitions they have contrived. Both of these processes gathered momentum in England during the second half of the sixteenth century as this paradoxical nation—both deeply parochial and peculiarly cosmopolitan—struggled to find its position in an unstable hierarchy of political, economic, and ideological relationships, and as an embryonic notion of "empire" gradually made its way from the pages of Spanish chronicles into the pages of English poetry, including those of *The Faerie Queene.*

How *did* the Spanish New World make its way into Spenser's poem? In the absence of precise biographical evidence, it is necessary to proceed by inference. Those Elizabethans who interested themselves in Spanish colonial accomplishments were small in number and spoke to a limited audience. Yet among the minority who cared about such matters were some of the most prominent lights in the state: Sir Walter Ralegh, William and Robert Cecil, Sir Francis Walsingham, Sir Christopher Hatton, the Sidneys, the Howards, and the Dudleys. Spenser may well be considered part of their audience, since during his stints in England he was on the periphery of the circles in which these individuals moved. This community of "projectors" was served in its interests by a small but dedicated group of translators and editors, among them Richard Eden, who Englished large sections of Martyr's *Decades de Orbe Novo* as well as sizeable fragments from Gonzalo Fernandez de Oviédo and a number of other continental authors; Thomas Nicholas, whose prolific output included Francisco López de Gómara's *Istoria de las Indias* and Augustín de Zárate's *Historia del descubrimiento y conquista del Perú;* the anonymous M. M. S., who offered *The Spanish Colonie,* a translation of a French version of Las Casas's *Brevissima Relación;* and, most importantly, Richard Hakluyt the Younger, whose *The Principall Navigations, Voiages and Discoveries of the English Nation* stands as one of the great literary achievements of the Elizabethan era.[9] Through these writers, a reader like Spenser—and one has to assume that he was a very energetic reader—would have had access to many, though certainly not all, of the major documents of the Spanish conquest in its various phases.[10]

Then there is the interesting question of Spenser's possible access during the late 1570s to one of the greatest archives in northern Europe of material related to colonization. This was the private library of the eclectic mathematician, occultist, cosmographer, court philosopher, and "projector" extraordinaire John Dee. A favorite of Elizabeth, a neighbor of Walsingham, and a former tutor of Leicester, Dee moved easily throughout the political, literary, and scientific estates in English society; in all of his activities he showed himself a zealous promoter of and apologist for efforts to find both a northwest and a northeast passage to Cathay. His main contribution to the expansionist aspirations of his peers was, however, his acquisition of useful books. Dee assembled an amazingly comprehensive collection of scientific and geographical texts; he also cataloged his collection. Fortunately, this catalog has survived in manuscript. Among Dee's holdings were two editions of Columbus's letters and a copy of *Historia del mondo nuovo* that belonged to Columbus's son Ferdinand; five editions of Vespucci's letters; four editions of Cortés's letters to Charles V; two editions of Martyr's *Decades,* as well as Eden's translation; Gómara's *Istoria;* chronicles of the Peruvian conquest by Zárate, Pedro de Cieza de León and Francisco de Xerez (three editions of the latter); two editions of Oviédo's *Historia general y natural de las Indias;* and the French translation of Las Casas's *Brevissima Relación.* In addition to these were copies of Jacques Cartier's *Navigation* (including John Florio's English translation), Martín Fernández de Enciso's *Suma de Geographica,* two editions of the narrative of Sir Martin Frobisher's first expedition in search of the Northwest Passage, Sebastian Munster's *Cosmographia,* Coronado's relation of his search for the seven cities of Cibola (in two editions), Jean de Lery's account of his voyage to Brazil, and Sir Humphrey Gilbert's pamphlet *A Discourse for a Discovery for a new Passage to Cataia.*[11] Ralegh, Sidney, and Leicester would have been able to use Dee's collection, as would Dee's longtime confidant Edward Dyer; Spenser's friendship with all of these men suggests that he could have used it too, though this cannot be proven. Certainly, Dee's library would have been the easiest place in England for Spenser to acquire the knowledge of the New World he displays in *The Faerie Queene.*

Spenser would have been influenced, too, by other kinds of cultural texts beyond the conventionally inscribed ones. A powerful example of such a text would be the mythos surrounding Sir Francis Drake, who was an anomaly among notable Elizabethans for having done much while writing relatively little.[12] The reports of his raids against the Spanish at Nombre de Dios in 1572, his circumnavigation of the globe from 1577 to 1580, and his crushing surprise attack on Spanish shipping at Cadiz in 1587 made him a daunting figure in the English popular imagination before the Armada ever set sail. He was a privateer and an opportunist, yet all of his actions appeared—at least in the secondhand accounts of those actions—to be deeds of loyal service to the Queen. Operating on a foggy frontier between legitimate naval activity and naked piracy, he was an enigmatic, if inspirational, figure.

Some of the more public Elizabethan personalities sought to present themselves as articulate versions of Drake, or as vehicles of a popular conception of his character. This seems to be the case with two of Spenser's important associates, Ralegh and Sidney. Ralegh, whose achievements as a sailor and military commander were decidedly inferior to Drake's, attempted to cast himself in his *Discoverie of the Large, Rich, and Bewtiful Empire of Guiana* as a leader in Drake's mold: disciplined, courageous, humane, patriotic, fiercely opposed to Spanish villainy, rapid in decision, and bold in execution.[13] Ralegh's claims for himself in the *Discoverie* (published in 1596, the year of Drake's death and the expanded edition of *The Faerie Queene*) cannot wholly conceal the sad truth that his search for El Dorado along the Orinoco River delta was a failure in nearly every respect. Ralegh is compelling as a tragic representative of the tensions in Elizabethan culture, but his heroic reputation was (and is) as much artifice as fact, and the artifice depended on the precedents set by Drake.[14]

Sidney could perhaps lay better claim than Ralegh to an independent greatness; yet he too was conscious of Drake as a powerful model. According to Greville, Sidney planned on shipping with Drake to the West Indies in 1585, but after a number of delays he was finally prevented from doing so by order of the Queen.[15] Greville later describes Sidney's ideas for weakening Spanish power, one of which was to "by the example of *Drake,* a mean born subject of the Crown of *England,* invade, possess, & inhabit some well chosen havens in *Peru, Mexico,* or both."[16] The unavoidable implication of Drake's being termed "a mean born subject" is that the noble Sidney would somehow improve on Drake's laudable record if only given the opportunity. Indeed, Greville goes on to describe Sidney as a kind of ideal Drake figure:

> To Martial men he opened wide the door of sea and land, for fame and conquest. To the nobly ambitious the far stage of *America,* to win honor in. To the Religious divines, besides a new Apostolical calling of the last heathen to the Christian faith, a large field of reducing poor Christians, mis-led by the Idolatry of *Rome,* to their mother *Primitive* Church. To the ingenuously industrious, variety of natural richesses, for new mysteries, and manufactures to work upon. To the Merchant, with a simple people, a fertile, and unexhausted earth. To the fortune-bound, liberty. To the curious, a fruitful womb of innovation.[17]

What lends this passage a certain strangeness is that Sidney in fact did none of these things in any practical sense. Some of this misrepresentation can be attributed to the pervasive tone of panegyric in Greville's *Life.* However, since Greville had long been acquainted with Sidney, the reader might well conclude that Sidney cultivated a likeness to Drake in both his private and public personas. Such indications in the lives of both Sidney and Ralegh suggest that Drake's importance lay not only with his actual service to England, but also with a notion of that service as a model of individual conduct for adventurers with higher public profiles than Drake's own—a model that in the right hands could be given an even more idealized and artful cast. It is the symbolism of Drake's character—the kind of behavior he was thought to embody, not Drake the historical personage—that finally makes its presence felt in Spenser's construction of the colonial gentleman in the 1590 edition of *The Faerie Queene.*

Texts and people were the concrete influences on Spenser's notion of colonial conduct, but a fuller outline needs to consider the more abstract ideological imperatives affecting his poetry. This raises the problem of what

Spenser, as a "poet of empire," would have understood "empire" to mean in the English context of the 1570s and 1580s. While Dee sometimes described his now-fragmentary treatise *General and Rare Memorials pertayning to the perfect Arte of Nauvigation* (1577) under the alternate title "The Imperium Brytannicum,"[18] the word "empire" was an uncommon one in the Elizabethan vocabulary. When it appeared in the promotional literature of the time, it usually referred to the Holy Roman Empire, and by extension to the Spanish line of the Habsburg dynasty, which under Philip II was a far more imperialist line than its Austrian counterpart under his uncle Ferdinand's progeny. It is true that in 1533 Henry VIII (or, perhaps more accurately, Thomas Cromwell) had announced "This Realm of England is an Empire" at the beginning of the Act in Restraint of Appeals, but the king's notion of empire was distinctly insular in character. Through an appeal to Roman law he attempted to distance the realm from the universal claims of the papacy and its nominal protector, Charles V, thus promoting the new order of state and church in England.[19]

English promoters in the later years of Elizabeth's reign drew upon a far more generalized and open-ended definition of the *imperium;* even so, they approached the concept or possibility of an overseas empire with a fair degree of caution, emphasizing instead the potential for religious conversion and improvement of trade (particularly of the market for English woolens) in the Americas. In his 1585 pamphlet on behalf of the Virginia enterprise, Richard Hakluyt the Elder devised the following table of priorities for colonization:

The ends of this voyage are these:	1. To plant Christian religion.	Or, to do all three.
	2. To traffic.	
	3. To conquer.	

He then went on to evaluate this table: "To plant christian religion, without conquest, will be hard. Traffic followeth conquest; conquest is not easy. Traffic without conquest seemeth possible, and not uneasy. What is to be done, is the question."[20] The answer, for Hakluyt, was to avoid a military occupation of Virginia if at all possible, "for a gentle course without cruelty and tyranny best answereth the profession of a Christian, best planteth Christian religion; maketh our seating most void of blood, most profitable in trade of merchandise, most firm and stable, and least subject to remove by practice of enemies."[21] This reads as a prescription not so much for empire as for a successful trading economy. At the same time, the elder Hakluyt's pamphlet does not really dismiss the notion of conquest; it holds the notion in suspense, as it were. Where other references to an Elizabethan empire occur, they are often left dangling in this fashion. Hakluyt the Younger, addressing Ralegh in the dedicatory epistle to his Latin translation of Martyr's *Decades* (1587), spoke of Elizabeth as an empress but in a carefully curtailed way: "Restat adhuc tibi novae terrae, regna amplissima, gentes ignotae, tibi, inqua, restant adhuc detegendae, sceptróq; serenissimae Elisabetae nostrae, maris Oceani, Hispano confitente, imperatricis, foelicibus tuis & armis & ausis brevi & facili negocio subigendae."[22] Despite the prowess of Elizabethan sailors, the English never controlled the Atlantic during the sixteenth century, and at any rate the "Ocean Sea" could only be considered an empire in a very abstract sense. Hakluyt gestures here toward an empire along more traditional lines, but the claim to empire is never stated outright. Likewise, in Book 1 of George Puttenham's *The Arte of English Poesie* (1589), in chapter 23 on "The Form of Poetical Rejoicings," Puttenham alludes to imperial hopes while discussing public celebrations in commemoration of "the public peace of a country, the greatest of any civil good; and wherein your Majesty . . . have showed your self to all the world . . . above all other Princes of Christendom, not only fortunate, but also most sufficient, virtuous, and worthy of Empire."[23] To be worthy of empire is not yet to possess one, whatever the future may hold. Thus Puttenham's praise partakes of the same ambiguity on the subject that appears in the work of the two Hakluyts. This is not to suggest that all writers during the period were so politic. For Ralegh, Guiana was not merely a landfall but an "Empire," and Spenser dedicated the 1596 edition of *The Faerie Queene* to Elizabeth as "THE MOST HIGH, MIGHTIE And MAGNIFICENT EMPRESSE . . . OF ENGLAND FRAVNCE AND IRELAND AND OF VIRGINIA." The high idealism of this dedication is suggested by the inclusion of Virginia in the old formula of the monarch's "empire."

The tentativeness with which the English embraced the vision of the "Imperium Brytannicum" probably had many sources, but an important one might be found in the period immediately preceding Elizabeth's reign, when relations between England and Spain were actually quite close. Beginning with the marriage of Prince Arthur to Catherine of Aragon as part of Henry VII's efforts at international diplomacy, the intimacies between the House of Tudor and the Spanish monarchy culminated in the marriage of Mary I to Prince Philip, soon to be Philip II—a marriage that put England on the brink of becoming part of the nominal empire of the Habsburgs. The rule of Philip and Mary proved to be an exceptionally difficult period for England, yet we find in the literature of the time a perhaps natural effort, given Philip's position, to praise the achievements of Spain as it stood at the height of its prosperity. In the preface "to the reader" in his 1555 translation of the *Decades,* Eden places the Spaniards above the Old Testament patriarchs in their manner of pacifying the savages of the wilderness:

> Moses as the minister of the law of wrath and bondage given in fire and tempests, was commanded in his wars to save neither man, woman, nor child, and yet brought no commodity to the nations whom he overcame and possessed their lands. But the Spaniards as the ministers of grace and liberty, brought unto

> these new gentiles the victory of Christs death whereby they being subdued with the worldly sword, are now made free from the bondage of Satans tyranny.[24]

Eden follows this celebration of the Spaniards' conduct with an admonition to the English for failing to come up to the same standard:

> How much therefore is it to be lamented, and how greatly doth it sound to the reproach of all Christendom, and especially such as dwell nearest to these lands (as we do) being much nearer unto the same than are the Spaniards (as within xxv. days sailing and less) how much I say shall this sound unto our reproach and inexcusable slothfulness and negligence both before god and the world, that so large dominions of such tractable people and pure gentiles, not being hitherto corrupted with any false religion (and therefore the easier to be allured to embrace ours) are now known unto us, and that we have no respect neither for gods cause nor for our own commodity to attempt some voyages into these coasts, to do for our parts as the Spaniards have done for theirs, and ever like sheep to haunt one trade, and to do nothing worthy memory among men or thanks before god, who may worthily accuse us for the slackness of our duty toward him.[25]

Eden's conceptual model for English enterprise in the New World simply emulates or reflects Spanish activity. At a time when England was indeed perilously close to falling under Spanish dominion, Eden urged his countrymen to identify with the Spaniards. Paradoxically, he mingled this plea for identification with a claim of independence, describing his proposed English settlement in the New World as a self-sufficient project, undertaken by the nation in its own economic and religious interests. Eden was, after all, not suggesting that the English convert these "tractable people and pure gentiles" to the church of Rome. In making such remarks, Eden attempted to thread a rather difficult needle, and he paid a penalty for his efforts. Though the official reasons are obscure, contemporary records indicate that Thomas Watson, bishop of Lincoln, accused Eden of heresy, citing his recent book as evidence. The case was heard before Stephen Gardiner, bishop of Winchester, one of Mary's chief agents in the attempt to suppress Protestantism in England. Gardiner died before the case could reach a verdict, and Eden survived relatively unscathed, though he lost his employment in Prince Philip's English treasury.[26] Evidently, Eden managed to offend the very people he had praised so fulsomely in his preface. Nor was this praise very satisfactory from the English point of view: When Richard Willes published an expanded edition of the *Decades* in 1577, Eden's original preface was completely expunged.

Eden's difficulties serve to suggest the contradictions inherent in England's national life during the mid-sixteenth century and continuing well into Elizabeth's reign. On the one hand there is a desire to recast the nation in the Spanish mold, and on the other there is a burgeoning consciousness of England's separateness from its powerful continental neighbors. In the aftermath of the Marian persecutions, a majority of the English came to think of the nation as unequivocally Protestant in character. At the same time, English merchants were making incursions into markets previously monopolized by Portugal and Venice.[27] While Spain remained the dominant imperial model for the English, an awareness of England's distance—and growing ability to distance itself—from that model began to affect the tone of Elizabethan thought on the Americas. In the dedication to Walsingham that precedes Nicholas's 1578 translation of Gómara, Nicholas offers his readers another opportunity to see a reflection of themselves in the Spanish experience:

> [This history] is a Mirror and an excellent precedent, for all such as shall take in hand to govern new Discoveries: for here they shall behold, how Glory, Renown, and perfect Felicity, is not gotten but with great pains, travail, peril and danger of life: here shall they see the wisdom, courtesy, valor and policy of worthy Captains, yea and the faithful hearts which they ought to bear unto their Princes service: here also is described, how to use and correct the stubborn & mutinous persons, & in what order to exalt the good, stout and virtuous Soldiers, and chiefly, how to preserve and keep that beautiful Dame *Lady Victory* when she is obtained.[28]

Here the metaphor of reflection has taken quite a different meaning from the one it possessed in Eden's preface; the English reader gazes in the "Mirror" of the Spanish text and sees not a better Spaniard but a better Englishman, as becomes clear in a subsequent passage: "And where our Captain *Hernando Cortez* . . . hath deserved immortal fame, even so doubtless I hope, that within this happy Realm is now living a Gentleman, whose zeal of travail and valiant beginnings doth prognosticate great, marvellous, and happy success."[29] The identity of the "Gentleman" is unclear;[30] what *is* clear is that Nicholas expected some representative Englishman to demonstrate his equality with the archetypal conquistador in the immediate future. The same note is sounded in Stephen Gosson's contribution to Nicholas's text. The sixth and last stanza of his poem "in prayse of the Translator" runs as follows:

> Lo here the traveller, whose painful quill,
> So lively paints the Spanish Indies out,
> That English Gentlemen may view at will,

> The manly prowess of that gallant rout.
> And when the Spaniard vaunteth of his gold,
> Their own renown in him they may behold.[31]

In the concluding two lines, Gosson's English gentlemen see not only the reflection but also the cause of their fame: Spanish gold. This is not merely a display of Gosson's approval of privateering in the fashion of Drake and Hawkins; the poem advances a claim to right of possession—of Spain's "renown," if not of Spain's American wealth. That such a claim could be advanced, even in an insignificant piece of occasional poetry, marks a considerable progression since Eden's time in English confidence about the nation's prominence on the European stage, and about her influence over affairs on the Ocean Sea.

After Drake's successful circumnavigation, that confidence grew stronger still. In composing the preface to his translation of Zárate in 1581, Nicholas was less reserved than he had been three years previously: "I may at this day, God be praised, boldly write that, where the Spanish and Portingal Nations dare glory of their discoveries & navigations, with great commendations of their Captains, Colon, Vasco de Gama, Magalanez, Hernando Cortez: don Francisco Pisarro, & Don diego de Almagro. Now may our most gracious Queen, most justly compare withall the princes of the world, both for discovery & navigation."[32] Plainly, Nicholas placed a great deal of faith in Drake's activity as a panacea for England's deficiencies as an expansionist state, for at the time Drake's circumnavigation was almost the only achievement of "discovery & navigation" by which Elizabeth could "compare withall the princes of the world." Other writers placed their faith in higher sources. In his report of Gilbert's disastrous 1583 voyage to Newfoundland, included in the first edition of Hakluyt's *Principall Navigations* (1589), Edward Hayes began with a frank admission of English weaknesses ("Many voyages have been pretended, yet hitherto never any thoroughly accomplished by our nation, of exact discovery into the bowels of those main, ample, and vast countries, extended infinitely into the North"[33]) but went on to assert that the Spanish had themselves been weakened—at least as far as their efforts in North America were concerned—by God's own hand:

> whensoever . . . the Spaniards (very prosperous in all their Southern discoveries) did attempt any thing into *Florida* and those regions inclining towards the North: they proved most unhappy, and were at length discouraged utterly by the hard and lamentable success of many both religious and valiant in arms, endeavoring to bring those Northerly regions also under the Spanish jurisdiction: as if God had prescribed limits unto the Spanish nation which they might not exceed: as by their own gests recorded may be aptly gathered.[34]

Hayes's musings on predestination hint at the rising voice of Puritanism, more evident and strident here than in the texts examined previously in this chapter. But Hayes was writing in the 1580s, the decade in which England realized, for good or ill, its religious as well as political and economic independence from Spain. By 1585 Elizabeth was at war with Spain in the Netherlands, largely over issues of religious autonomy; and England's sense of its own integrity as a state came to a triumphant head with the defeat of the Armada in 1588, an event that convinced most Englishmen that God had indeed "prescribed limits unto the Spanish nation."[35]

Hayes's remarks above also suggest how inwardly complex English relations with Spain actually were. For his judgment of the Spaniards, Hayes depended on "their own gests recorded"—on Spanish texts. He was not alone in this dependency, as should now be clear, and his use of the Spanish literary legacy points up a crucial paradox—perhaps *the* crucial paradox—in the development of the concept of an Anglo-American empire during the middle years of Elizabeth's reign. England's heightened consciousness of nationhood in these years and the abundance of patriotic effusions that sprang from that consciousness were both due in part to the presumption that England was not like Spain, that its identity as a state was utterly separate from Spain's and from Spanish interests. At the same time, Spain remained for England the exemplar of a particular style and magnitude of imperial government through its comprehensive recording of its own activities. As part of his reading of Hakluyt's *Principall Navigations,* Richard Helgerson remarks aptly that "The ideological relation of England to Spain was . . . complicated by an awkward mix of similarity and difference. . . . England could . . . feel comfortable neither in a complete repudiation of the Spanish model nor in an unqualified imitation of it."[36] In a striking way, the Tudor proponents of the American venture exploited the matter of Spain to declare their independence from the essence of Spain. This unacknowledged debt to the Spaniards helps account for the reticence of Elizabethan writers in talking about "empire." Before such a term could be applied to England, there had to be some assurance that England stood apart from its imperial prototype. This assurance was long in coming, not only because the rift with Spain took the form of a gradual deterioration rather than a sudden rupture, but also because England's literary and intellectual connections with Spain declined much more slowly than its political and religious connections.

The naval achievements of the 1580s enabled the English to view their nation as a power among powers and allowed Spenser and Ralegh to speak of an "empress" and an "empire" with a panache that earlier writers could not muster. Yet amidst the braggadocio that begins to creep into both poetic and promotional literature in the 1580s and 1590s, one finds a nagging sense of hollowness and uncertainty related to the manner in which

England set about the work of empire in North America, an awareness that an English landscape was being superimposed on a conceptual grid provided by the Spaniards. Such promoters were eager to find the attributes of empire that could properly be ascribed to England. But because the identity of the English empire was so closely bound to a negative model, and because English notions of what could be accomplished in North America were so intimately linked in the 1580s with a desire to do things differently from the Spaniards, the writers and thinkers who drafted schemes for the New World enterprise were faced with a conceptual void and the difficult task of filling it. The concept of Elizabeth's monarchy over Virginia and parts beyond could not subsist simply on ideas about what it was not; in the absence of concrete evidence of England's empire (such as a successful American settlement), that concept became largely a matter of ingenious use of the available materials, a pastiche of Spanish descriptions of the New World inverted to suit English needs. Elizabethan promoters, inventive as they were, held no real advantage over Elizabethan poets at devising this sort of literary construction.

In 1590, in the midst of the peculiar ferment over England's prospects in Virginia among the courtiers, soldiers, merchants, and scholars of London, Bristol, Oxford, and Cambridge, Spenser published the first three books of *The Faerie Queene*. Though in political terms Spenser is a minor figure in the history of England's ascendancy under Elizabeth to prominence in the commercial, maritime, and religious affairs of Europe, *The Faerie Queene* nonetheless stands as a lasting monument to that ascendancy. Spenser used his poem to broach the same matters, large and small, that concerned the great historical actors of his generation; the poem remains, after Shakespeare's collected plays, the most majestic repository of Elizabethan culture in all its glories and failings. It is as part of his effort to comprehend and reify the Tudor polity in all its fullness that Spenser considers, in the poem's second book, and in a manner that can best (and perhaps only) be described as "Spenserian," the glories and failings of England's nascent imperialism.

CHAPTER 2

Bloody Hands and "Puissant Kings": Guyon's Mission

Since the question of how English identity is to be advanced or preserved when put to the test is a crucial theme in *The Faerie Queene,* it is fitting that Guyon, the knight of temperance and the hero of Book 2, participates in some sort of orientation on this same question—the poetic equivalent of passing through customs with one's papers in order—before continuing his quest. Such an orientation actually occurs twice in Book 2, at the two castles that Guyon visits on his journey.

At the first castle, Guyon responds to Medina's request "To tell from whence he came through ieopardie, / And whither now on new aduenture bound" (2.2.39.5–6) by, in effect, producing a letter of conduct from the highest possible authority. He performs a eulogy to

> that great Queene,
> Great and most glorious virgin Queene aliue
> That with her soueraigne powre, and scepter shene
> All Faery lond does peaceably sustene.
> In widest Ocean she her throne does reare,
> That ouer all the earth it may be seene;
> As morning sunne her beames dispredden cleare,
> And in her face faire peace, and mercy doth appeare.
> (2.40.2–9)

The double repetition of "great" and "all," the juxtaposition of "Ocean" (not merely "wide" but "widest") and "earth" in consecutive lines, and the simile of the "morning Sunne" casting its rays endlessly toward the west all contribute here to a sense of Gloriana as a physically encompassing force that crosses conventional boundaries. Of course, the boundaries crossed here belong to Fairyland, a landscape over which the poet himself sees farthest. Spenser expands on the general theme in the next three stanzas, employing three variations on the phrase "all the world":

> And all that else this worlds enclosure bace
> Hath great or glorious in mortall eye,
> Adornes the person of her Maiestie. . . .
> (2.41.3–5)

> . . . on me she deigned to bestowe
> Order of *Maydenhead,* the most renownd,
> That may this day in all the world be found. . . .
> (2.42.3–5)

> My soueraine,
> Whose glory is in gracious deeds, and ioyes
> Throughout the world her mercy to maintaine,
> Eftsoones deuisd redresse. . . .
> (2.43.5–8)

Guyon concludes his panegyric with the remark that he has traveled for three months since leaving Gloriana's court (2.44.1–3)—the implication being that much distance has been covered. Notably, stanzas 40–43 each repeat the word "all" twice, and "all" is echoed in words like "mortall" (2.41.4) and "royall" (2.44.4). Plainly, Guyon serves a universal ruler, or at least one who appears (from the poets speculative vantage) to have the capacity to rule universally.

However, within this frame of allegorical Elizabetheism—Gloriana is actually described as an "Idole" here (2.41.9)—there are some puzzling features from either a narrative or a historical perspective. First, there is the problem of narrative causation: Who initiates Guyon's quest? Not, it seems certain, Guyon himself, for he is assigned the task by Gloriana: "Me all vnfit for so great purpose she employes" (2.43.9). Spenser uses "all" in this line to emphasize abasement rather than dominion. Yet the Fairy Queen has "deuisd redresse" in response to a request from the Palmer (2.43.1–5). So in a sense the Palmer is originally responsible for Guyon's journey in search of Acrasia. That does not fit smoothly, though, with the account of the journey presented in the first canto. Nowhere during their encounter with Amavia do Guyon and the Palmer indicate that they are already familiar with Acrasia's deeds, that much of Amavia's story is, strictly speaking, redundant information. Guyon seems at the time to have little sense of a larger quest other than what he generates on the spot: "all I seeke, is but to haue redrest / The bitter pangs, that doth your heart infest" (1.48.4–5). Here, too, Guyon offers redress in "global" terms: "That I may cast to *compasse* your reliefe" (1.48.8; my emphasis). It is the terrible scene presented in stanzas 39–41 that leads Guyon to take a "sacred vow" of "dew vengeance" (2.60.9, 2.61.7) and directs his course for the remainder of the book.

> . . . he rusht into the thicke
> And soone arriued, where that sad pourtraict
> Of death and dolour lay, halfe dead, halfe quicke,
> In whose white alabaster brest did sticke
> A cruell knife, that made a griesly wound,
> From which forth gusht a streme of gorebloud thick,
> That all her goodly garments staind around,
> And into a deepe sanguine dide the grassie ground.
>
> Pitifull spectacle of deadly smart,
> Beside a bubbling fountaine low she lay,
> Which she increased with her bleeding hart,
> And the cleane waues with purple gore did ray;
> Als in her lap a louely babe did play
> His cruell sport, in stead of sorrow dew;
> For in her streaming blood he did embay
> His litle hands, and tender ioynts embrew;
> Pitifull spectacle, as euer eye did view.
>
> (1.39–40)

The spectacle of the consequences of Amavia's intemperance inspires in the reader not only pity but also revulsion, with the passage's clinical emphasis on "gorebloud thick" and "purple gore," its alignment of "bubbling" and "bleeding" to create an impression of blood gushing fountain-like from Amavia's wound, and its bizarre image of the infant bathing its hands in this gruesome "fountain." This revulsion (basically horror at Acrasia's destructive influence on human life) would seem to be the immediate, visceral cause of Guyon's quest—not an assignment from the Fairy Queen or a plea from the Palmer.

What I want to point out here is the strangely passive and circumstantial quality of the quest in Book 2, which as Spenser has plotted it seems to originate outside the system devised in the letter to Ralegh ("that the Faery Queene kept her Annuall feaste xii. dayes, vppon which xii. seuerall dayes, the occasions of the xii. seuerall aduentures hapned, which being vndertaken by xii. seuerall knights, are in these xii. books seuerally handled and discoursed.") The premise for Book 2 that Spenser outlines in the letter actually contradicts the sequence of events in the book itself: "The second day ther came in a Palmer bearing an Infant with bloody hands, whose Parents he complained to haue bene slayn by an Enchaunteresse called Acrasia."[1]

To say that Spenser changed his mind when he sat down to write Book 2, or that the letter to Ralegh should not be taken too seriously as a key to *The Faerie Queene,* is not to say enough on the problem. Spenser seems to be mystifying the causes of Guyon's activity, confusing the formal and efficient elements so as to deflect attention from any single point of origin. Significantly, Guyon's quest does not originate within the order to which he belongs, but in an event that occurs without reference to that order.[2] The quest comes across as less than fully motivated; in some sense it just "happens."

From this state of affairs emerges the historical puzzlement represented in stanza 40 of canto 2. Given the inevitable identification between the Fairy Queen and Elizabeth, one might be led to ask what sort of peace and mercy, at this point in time, Elizabeth's "morning Sunne" had spread across the "widest Ocean" and "ouer all the earth." Elizabeth's England is a much more circumscribed realm than Gloriana's Fairyland. And if Guyon is supposed to figure in the poem as a representative of that "soveraigne powre" in the larger world, what can really be said about his accomplishments when they are translated into historical terms? Spenser implies the existence of grand imperial projects already implemented, but the reality in 1590, as he was surely aware, was far otherwise. The cryptic features of Guyon's quest point to Spenser's basic coyness in describing a universalized (and historicized) Elizabethan "quest." He might be said to follow the advice that greets Britomart in Book 3 from above the last iron door in Busyrane's castle: "*Be not too bold*" (3.11.54.8).

Consciousness of Elizabethan England's impoverished situation as an "empire"—impoverished especially in comparison with the vast military, ecclesiastical, and bureaucratic networks already established by Spain in the New World—is not only available in retrospect. The great upswing in the writing, translation, and publication of geographical literature in England from the 1570s onward reflects a growing popular sense of the deficiency, as well as the promise, of British colonial activity. "Peace" and "mercy" might be provisionally meaningful as virtues practiced by Elizabeth in her domestic policy, but with regard to expansion beyond the British Isles these words were still merely variables without values attached to them. If they were to be defined specifically as English virtues, the definition would have to be in contrast to "peace" and "mercy" as such virtues were practiced by the archetype of successful overseas empire in the sixteenth century.

While the English were deeply ambivalent about Spain's accomplishments in the New World under Charles V and Philip II, I think it is safe to say that during the stressful times immediately before and after the failure of the Armada, the average citizen's opinion would have been that the Spaniards did not know the meaning of either peace or mercy. By 1588 Spain had become the antithesis of all things English, and the unstated goal of English international policy was to strive to be whatever the Spaniards were not. The value of peace and mercy lay less with their intrinsic value than with the idea that these terms were, so to speak, outside of Spanish territory.

The Amavia episode itself suggests the immanence of such negative historical exempla, as well as their power in shaping Guyon's purposes. One effect of Amavia's "Pitifull spectacle" is to evoke associations with ritualized human sacrifice; the image of Ruddymane bathing his hands in Amavia's wound makes such associations nearly unavoidable. Of the many sensational aspects of New World cultural life that entered the European imagination in the sixteenth century, human sacrifice—and its remarkable ubiquity among the tribes of Central America—was probably the most shocking. Martyr, recounting Juan de Grijalva's expedition to Cozumel in 1518, describes with combined horror and fascination the discovery of the Mayans' version of the practice:

> oh abominable cruelty: oh most corrupted minds of men, and devilish impiety? Let every godly man close the mouth of his stomach lest he be disturbed. They offer children of both kinds to their Idols of marble and earth. . . . Let us now declare with what ceremonies they sacrifice the blood of these poor wretches. They cut not their throats, but open the very breasts of these silly souls and take out their hearts yet panting, with the hot blood whereof, they anoint the lips of their Idols, and suffer the residue to fall into the sink [at the base of the idol]. . . . They [Grijalva's men] found a stream of congealed blood as though it had run from a butchery.[3]

This complex of disturbing images, or one very much like it, is refracted into several episodes of *The Faerie Queene:* the description in Book 1 of Orgoglio's dungeon "filthy . . . With bloud of guiltlesse babes, and innocents trew" (1.8.35.5–6); the spectacle in Book 3 of Amoret's heart "drawne forth, and in siluer basin layd, / Quite through transfixed with a deadly dart, / And in her bloud yet steeming fresh embayd" (3.12.21.2–4); and, most explicitly, the moment in Book 6 when Serena nearly meets her doom at the hands of the cannibal priest (6.8.45). In this last instance, however, the "decorum" of sacrifice—practiced as it is by a "salvage nation" (6.8.35.2), of which such things might be expected—might be more tolerable to the sixteenth-century reader than the scenes in Orgoglio's dungeon and Busyrane's castle, for which ignorance could not afford an excuse. Sacrifice perpetrated under the auspices of "civilization" (that is, civilization according to European conventions) becomes doubly appalling. This idea alone may be sufficient to explain the strong impulse among writers of the late sixteenth century to "borrow" the practice of sacrifice from the Amerindians and bestow it on the Spaniards.

In the passage from Martyr, sacrifice is associated with other, more familiar crimes against God and man: infanticide and idolatry. Ironically enough, associations such as these helped to transform the Spaniards as they appeared in polemical literature (mainly but not exclusively in the Protestant varieties of such literature) from opponents of sacrifice to participants in it. The old Reformation truism that Roman Catholics were really idol-worshippers; the powerful exemplum of Herod's slaughter of the innocents; the historical residua of the Saint Bartholomew's Day massacre, the Duke of Alba's despotism in the Netherlands, and the assassination of the Prince of Orange; the Black Legend of Spanish atrocities in the New World that collected around the writings of Las Casas;[4] all of these things contributed to an image of the Spaniards as corrupted by their experience in and of the New World and implicated in the worst sins of the barbarians they had supposedly redeemed from perdition.

Las Casas himself, as is so often the case, provides one of the most grotesque versions of "the Spanish character" in the New World. In the *Brevissima Relación,* he relates an episode in which a handful of Franciscan monks, engaged in a successful effort to convert the natives of the Yucatan to Christianity, are intruded upon by a band of idol-toting slavers who proceed in brutal fashion to restore the old religion. The anonymous translator of *The Spanish Colonie* presents the episode as follows:

> The captain of those thirty Spaniards, called unto him a Lord of the country . . . and commandeth him to take those idols, and to disperse them throughout all his country, selling every idol for an Indian man, or an Indian woman, to make slaves of them, with threatening them, that if he did not do it, he would bid them battle. That said Lord being forced by fear, distributed those Idols throughout all the country, and commanded all his subjects, that they should take them to adore them, and that they should return in exchange of that ware Indies and Indisses to make slaves of. The Indians being afraid, those which had two children, gave him one, and he that had three gave him two. This was the end of this sacriligious traffic: and thus was this Lord or Cacique, fain to content these Spaniards: I say not Christians.[5]

Here slavery stands in for human sacrifice, though the end result is the same: the innocents are sent to the slaughter as part of a blasphemous exchange, made still more blasphemous by the fact that its victims are literally forced back into error by the alleged bearers of truth.

No explicit connection can be drawn between events in Las Casas's Yucatan and those in Spenser's Fairyland; even so, the passage above throws an interesting, if indirect, light on several aspects of Amavia's narrative. In attempting to rescue Mordant from Acrasia and the Bower of Bliss, Amavia disguises herself "in Palmers weed" (1.52.8); she approaches the Bower in the role of a religious pilgrim and, by extension, under the appearance of ecclesiastical authority. She travels a great distance—"three quarters" of the lunar year (1.53.2)—toward the Bower before giving birth to Ruddymane in a wild and foreign landscape: "The woods, the Nymphes, my bowres, my midwiues weare, / Hard helpe at need" (1.53.7–8). Out of necessity Amavia must make concessions to the wilderness she has come to inhabit. When she at last finds Mordant, he is "so transformed from his former skill / That me he knew not, neither his own ill" (1.54.4–5). The suggestion here is that Mordant fails to recognize not only Amavia herself but also Amavia dressed as a palmer, which in turn suggests that Mordant no longer recognizes what palmers are or what they represent. The "deliueraunce" (1.54.9) of Mordant is forestalled when he succumbs to the charmed cup, leaving Amavia to die with the words "I wretch" on her lips.

Amavia's tale presents the reader with a sequence of familiar archetypes from the history (both actual and imagined) of the conquest of the New World. A soldier enters that world first, "his puissant force to proue" (1.50.7). He falls prey to vices that he once resisted and at last no longer knows "his owne ill." He is closely followed—one might even say pursued—by a figure in religious garb who strives to right the balance but who, like the soldier, comes to "dyen bad" (1.59.9). The knight and the "palmer" leave the next generation without guidance in an unfamiliar land, "carelesse of . . . woe, or innocent / Of that was doen" (2.1.7–8), bearing a "bloudguiltiness" (2.4.5) more immediate to the eye than original sin. The hope held out for Ruddymane is that he will grow up to repudiate his tragic heritage; in leaving him with Medina, Guyon not only charges her "In vertuous lore to traine his tender youth," but also orders that he be "taught, / T'auenge his Parents death on them, that had it wrought" (3.2.4, 8–9). By learning to negate the acts of his forebears, Ruddymane achieves his own redemption, and it is Guyon who proposes the course of this negation.

Thus when Guyon and the Palmer leave Medina's castle, the task before them is not to avenge what has already transpired but to undo it. The roles of the knight and the pilgrim have already been established by Mordant and Amavia; Guyon and the Palmer must play the same roles, but the moral imperative of Book 2 is that they not play them in the same way. It is appropriate that Guyon, viewing Mordant's and Amavia's remains, should define temperance as a virtue founded on negation: "*Neither* to melt in pleasures whot desire, / *Nor* fry in hartlesse griefe and dolefull teene" (1.58.3–4; my emphasis).

The second and more obvious instance of ideological orientation in Book 2 occurs in canto 10, where Arthur and Guyon receive a political and historical education to complement the moral education they received in canto 9. Guyon's reading of the "*Antiquitie* of *Faerie* lond" takes up only a brief space in the canto, yet within this space Spenser is at pains to portray Fairyland as a full-fledged empire rather than a mere kingdom:

> Of these a mightie people shortly grew,
> And puissaunt kings, which all the world warrayd,
> And to them selues all Nations did subdew:
> The first and eldest, which that scepter swayd,
> Was *Elfin;* him all *India* obayd,
> And all that now *America* men call.
> (10.72.1–6)[6]

The region over which Elfin once ruled is "now" called America; the "now" implies that America belongs among the "antiquities, which no body can know" in the proem. As part of Spenser's ahistorical history, America is also part of England's promised future. Spenser's parallel placement of India and America recalls the confusion between East and West for which Columbus continues to provide the exemplary instance; at the same time this parallelism suggests the encompassing nature of the Fairy empire, an empire so well-traversed that its mysterious regions can now be clearly distinguished from each other. A later Fairy king is notable for having closed a distance over water, presumably extending his empire in the process: "He built by art vpon the glassy See / A bridge of bras, whose sound heauens thunder seem'd to bee" (10.73.8–9).

The book that Guyon reads offers only the fact of empire without a comparable explanation for its existence; this last belongs to the longer and far more interesting narrative that presents itself to Arthur as, among other things, a collection of precedents relevant to his role, both inside and outside of time, as the perfect Briton. While the *Antiquitie* of Fairyland anticipates the future in its version of the past, the *Briton moniments* draws powerful connections between the distant past and the immediate historical (read colonial) present.[7] Spenser may draw his material from the chronicles, but the placement of this material among so many overt and covert allusions to overseas empire evokes the notion of a new English world being projected onto the face of ancient Britain.

This effect arises partly out of the fact that the discourse of the Fairy "past" and that of the Spenserian historical present are inflected so similarly. For example, prior to recounting the book's contents, Spenser pays tribute to Elizabeth in terms that recall those employed by Guyon to Medina in praise of the Fairy Queen:

> Ne vnder Sunne, that shines so wide and faire,
> Whence all that liues, does borrow life and light,
> Liues ought, that to her linage may compaire,
> Which though from earth it be deriued right,
> Yet doth it selfe stretch forth to heauens hight,

> And all the world with wonder ouerspred;
> A labour huge, exceeding farre my might.
> (10.2.1–7)

This stanza shares the same sort of "global" vocabulary that appeared in stanza 40 of canto 2. There are other echoes of the earlier sequence at Medina's castle: repetitions of and assonances with the word "all," and variations on the adverb "farre" in four consecutive stanzas ("all earthly Princes she doth farre surmount" [10.1.9]; "A labour huge exceeding farre my might" [10.2.7]; "Thy name, O soueraign Queene, to blazon farre away" [10.3.9]; "that royall mace, / . . . to thee descended farre / From mightie kinges and conquerours in warre" [10.4.3–5]).

Having established that Elizabeth, like the Fairy Queen, is a world-embracing, distance-closing figure, Spenser proceeds to find the source of her actions in Britain's own geographical history: "The land, which warlike Britons now possesse, / And therein haue their mightie empire raysd, / In antique times was saluage wildernesse, / Vnpeopled, vnmanur'd, vnprou'd, vnpraysd" (10.5.1–4).[8] Here, Britain is like Peru and Virginia in the proem, "great Regions" that "were, when no man did them know." The promise of Britain is brought to light by a "venterous Mariner," who opens the way for the land's exploitation by other men: "later day / Finding in it fit ports for fishers trade, / Gan more the same frequent, and further to inuade" (10.6.2, 7–9). There is, however, an obstacle to the fulfillment of Britain's promise:

> But farre in land a saluage nation dwelt,
> Of hideous Giants, and halfe beastly men,
> That neuer tasted grace, nor goodnesse felt,
> But like wild beasts lurking in loathsome den,
> And flying fast as Roebucke through the fen,
> All naked without shame, or care of cold,
> By hunting and by spoiling liued then.
> (10.7.1–7)

The description of the "halfe beastly men" who lurk "like wild beasts" anticipates both Maleger's bestial army (11.8) and the men transformed into "seeming beasts" by Acrasia (12.85.1).

Spenser seems to convey his awareness of the idea, current around the time of *The Faerie Queene*'s first publication, that the "Giants" of the distant British past and savages of the New World might be comparable. In the 1590 edition of Thomas Hariot's *A Briefe and True Report of the New Found Land of Virginia* from 1586, Theodore de Bry (probably with the help of John White, the original source of de Bry's illustrations) makes an explicit visual equation between the Algonquians and the Picts "which in the old time did habit one part of the great Britain."[9] The American

Indians are, like these giants, a "saluage nation" who have "never tasted grace" in the form of knowledge of Christ or of Christian values.

> They held this land, and with their filthinesse
> Polluted this same gentle soyle long time:
> That their owne mother loathd their beastlinesse,
> And gan abhorre her broods vnkindly crime,
> All were they borne of her owne natiue slime.
> (10.9.1–5)

This "beastlinesse," with its hints of incest and cannibalism, reaches its ordained end with the arrival of Brutus, who "them of their vniust possession depriu'd" (10.9.9). In order to gain possession, Brutus must first reduce the population of giants: "He fought great battels with his saluage fone; / In which he them defeated euermore, / And many Giants left on groning flore" (10.10.3–5). This is a rather peremptory justification for conquest in either past or present time; it is based on the idea that a native people can lose their right to the land they inhabit because they have not "tasted grace," and that this loss of right is providential.

Yet Spenser has carefully insinuated doubts about the true "natiuity" of this native people, calling into question their possession of the land: "whence they sprong, or how they were begot, / Vneath is to assure" (10.8.1–2). Spenser dismisses the tale of Dioclesian's daughters cohabiting with demons to spawn these giants, but he dwells on the story long enough to create a lingering impression that the giants are in some sense specious inhabitants—intruders, invaders—who must be displaced by more authentic inhabitants. Such an identification subverts the relationship between conqueror and conquered: the new invader becomes the true, divinely appointed native, while the native becomes a foreign presence that must be expelled. This identification opens the way for another consequence as well; natives can now be classed with other invaders of the "native" like the barbarians in stanza 15:

> Vntill a nation straung, with visage swart,
> And courage fierce, that all men did affray,
> Which through the world then swarmd in euery part,
> And ouerflow'd all countries farre away,
> Like *Noyes* great flood, with their importune sway,
> This land inuaded with like violence,
> And did themselues through all the North display.
> (15.1–7)

In terms of the chronicle history, this stanza refers to the Huns, but it is also possible to detect in it a passing allusion to a contemporary "nation straung, with visage swart," which had "ouerflow'd all countries farre away," which was on "display" throughout northern Europe in the 1570s and 1580s, and which was threatening to invade England "with like violence."

Later in the canto, Spenser makes an explicit reference to Spain: King Gurgunt "gaue to fugitiues of *Spayne*, / Whom he at sea found wandring from their wayes, / A seate in *Ireland* safely to remayne, / Which they should hold of him, as subject to *Britayne*" (10.41.6–9). In the context of the present discussion this passage takes on considerable interest, for Spenser indicates that many of the Irish "natives" are, genealogically speaking, foreigners living in the land by the good graces of a British monarch; as foreigners, they are subject to being dispossessed of what is not truly theirs.[10] Moreover, they are Spaniards—a significant identification at a time when a Spanish invasion of Ireland was thought to be imminent.[11]

What Spenser manages to propose in canto 10 is not just a general rationale for conquest, but also a specifically English one. In the case of Ireland, to displace the native population from its territory and to Anglicize the country was to thwart England's greatest external enemy at the same time. Thus, the fear that the plantation English would accommodate themselves to traditional Irish culture, as many settlers in fact did, can be viewed as having more than one level; an accommodation with Ireland was also, in more obscure ways, an accommodation with Spain.

The case could be extended throughout the Spanish sphere of influence. The colonial literature of the sixteenth century provides many instances in which the Spaniards are described as taking on the attributes of their native charges. The conqueror begins to merge with the conquered—a much more insidious form of conquest, in English eyes at least, than straightforward dispossession. If conqueror and conquered become indistinguishable, then this creates considerable impetus for the Elizabethan imperial effort, because to invade and displace the "halfe beastly men" of the "saluage wildernesse" is only to invade and displace one's own potential conqueror.

CHAPTER 3

Negating the Conqueror: Guyon as Anti-Conquistador

Book II is plain sailing.

C. S. Lewis, *The Allegory of Love*

My general argument in this chapter is that the meanings of Book 2 unfold completely if the moral geography and the physical geography are considered genuinely equivalent—one and the same thing, in effect. This argument may initially appear counterintuitive, since the landscape (and waterscape) through which Guyon travels on his quest is in many ways typical of Fairyland. It cannot be mapped in the ready way offered by the carefully limned worlds of the *Divine Comedy* or *Paradise Lost;* even its fixed landmarks tend toward abstraction. These landmarks have usually been interpreted as pointing toward the poem's moral allegory. In his great, if idiosyncratic, *English Literature in the Sixteenth Century,* C. S. Lewis concluded, perhaps too readily, that "in Faerie Land it is all quite simple. All the states [i.e., those 'states of the heart' that Lewis has already enumerated] become people or places in that country."[1] The "simple" equation of "states" with places in Book 2 has mostly prevailed during the more than forty years since Lewis's history was published, with some small adjustments.[2] Lewis's durable "projection" involves a mapping of metaphysical rather than physical features, with the landmarks in Fairyland mainly serving as signposts directing the reader to more rarefied regions of the heart and soul. Against this familiar articulation of the Spenserian landscape, I will maintain that Guyon's actions in that landscape do not just point allegorically to moral constructions; his actions *are* moral in something like a physically realistic sense and thus integral to Spenser's purposes in Book 2. Useful insights can be gained, then, by taking literally the poet's efforts to create a concrete geography within which the protagonist has to operate and make difficult decisions. In another great mapping of *The Faerie Queene,* James Nohrnberg observed that "Spenser conceives temperance as the virtue that secures the private integrity against an unruly physical environment."[3] I devote the first part of this chapter to investigating this physical environment on its own terms, then propose that the world the knight of temperance encounters is unruly because, among other reasons, it is a colonial world.

Michael Murrin has argued forcefully from the evidence in Book 2 that "Fairyland is India and America," and that "Faery is thus a dream of empire," embracing both East and West.[4] I want to refrain from making such bold claims (though I mainly agree with them) and proceed more modestly, considering the mundane details of the landscape of Book 2 apart from the question of specific locales. At an obvious level, Book 2 is an allegory of the physical and moral health of the body, as well as of the threats to such health; in Guyon, Spenser extols the virtue of temperance as a bulwark against both internal and external assaults on human well-being and right-doing. It is certainly possible to argue that the primary landscape in the book is the microcosm of the body. At the same time, Book 2 is the most nautical section of a poem saturated with nautical imagery; while much of this imagery reflects a highly conventional use of metaphor, it does have a cumulative effect. At the end of Book 1 Spenser anticipates the next book with an allusion to his "vessell" and "long voyage whereto she is bent" (1.12.42.4, 8). Midway through the first canto the Palmer admonishes Guyon in terms appropriate to seafaring: "God guide thee, *Guyon,* well to end thy warke, / And to the wished hauen bring thy weary barke" (2.1.32.8–9). Two stanzas later, Spenser writes that "*Guyon* forward gan his voyage make" (1.34.3). At the beginning of Book 3 Spenser sums up the preceding adventure with special attention to the long distances Guyon and Arthur traveled on their return: "Full many Countries they did overronne / From the vprising to the setting Sunne" (3.1.3.4–5).

Considerable attention has been paid over the years to Spenser's allegorical use of water, the sea, and the sea voyage in the narrative,[5] but a flatly literal reading of these features leads to the intriguing conclusion that Guyon spends more time on the water than any other major figure in *The Faerie Queene*—this despite Lewis's characterization of temperance as a "pedestrian" virtue that keeps Guyon "on foot."[6] The only other major contender is Florimell in Book 3; the "Fishers wandring bote," like Phaedria's in Book 2, goes "at will, withouten carde or sayle" (3.8.31.1, 2). Florimell's voyage, however, seems mainly to serve as a prelude for her descent with Proteus to the sea bottom, which parallels the wounded Marinell's descent with Cymoent and leads to the great convergence of "waters" in cantos 11 and 12 of Book 4. Marinell, the most obviously "seaworthy" of Spenser's knights, appears by the sea and under it, but never really *on* it.[7] Guyon, on the other hand, embarks four times: twice in canto 6, when Phaedria ferries him to her island in the Idle Lake and then ferries him once more from the island to the opposite shore after his fight with Cymocles; for the third time in canto 12, as he sails toward his confrontation with Acrasia; and finally, by inference, as he turns away from the Bower of Bliss at the end of that same canto.

The arrangement of land, shore, and water in Book 2 is not entirely consistent, nor is it likely that Spenser intended it to be. Even so, in the interests of delineating Guyon's status within the landscape, I will try to sketch a plausible map of that landscape. At its center is the Idle Lake, variously described as "a river" (6.2.4), a "wide Inland sea" (6.10.1), a "great lake" (6.11.4, 6.18.7), and a "deepe" or "perlous" ford (6.4.4, 6.19.9). Between the Idle Lake and the body of water over which Guyon must voyage to reach the Bower of Bliss is the region given over to Mammon and his cave. The other major landmarks on this map are the two castles, Medina's and Alma's, both of which front the water and serve as ports of departure for Guyon. Medina's castle is "Built on a rocke adjoyning to the seas" (2.12.7), perhaps the "seas" of the Idle Lake, though Spenser never makes this clear; at least the castle and the lake are not very distant from each other. The House of Alma is "plast / Foreby a river" (9.10.3–4), the river that Guyon and the Palmer use to reach the sea (11.4). Spenser identifies both castles with the western horizon and the idea of facing toward the west. Guyon finishes recounting the tale of Mordant and Amavia to Medina as "Orion, flying fast from hissing snake, / His flaming head did hasten for to steepe" (2.46.2–3). The counterpart to this setting star is the setting sun that marks the arrival of Arthur, Guyon, and the Palmer at Alma's doors: "And now faire *Phoebus* fan decline in hast / His weary wagon to the Westerne vale, / Whenas they spide a goodly castle" (9.10.1–3).

Though Spenser adeptly suggests the general contours of a landscape, he is evasive about details. For instance, the Idle Lake contains an island very much like the Bower of Bliss, and Phaedria appears in proximity to both places,[8] yet the Bower itself seems to be located elsewhere, in another sea of equally indeterminate dimensions. Cymocles, summoned by Atin from the Bower of Bliss (5.28ff.) in order to avenge Pyrocles' assumed death, must still cross over the Idle Lake in Phaedria's boat (6.4); Spenser neglects to describe how Atin and Cymocles traveled between the Bower and the mainland in the first place. Alma's castle, at which Guyon and Palmer take ship for the Bower, gives the impression of being a considerable distance from the Idle Lake. And even though Phaedria and her boat make an unexpected appearance on Guyon's itinerary in canto 12 (not the only unexpected appearance: in canto 6 Pyrocles manages to rejoin Atin on the other side of the Idle lake, evidently without ever leaving land [6.41ff.]), Spenser takes pains to distinguish the Idle Lake from the present hazardous seas (12.17).

The suggestion of different but somehow vaguely connected seas becomes less mystifying in the context of Amavia's description of the Bower of Bliss as "a wandring Island, that doth ronne / And stray in perilous gulfe" (1.51.5–6); the Bower has no precise orientation. For that matter, Phaedria's island floats on the Idle Lake without being anchored in one place (6.11.4). These wandering islands reflect not only the allegorical tenor of Book 2 and the influence on Spenser of epic and romance conventions involving similar islands, but also the speculative nature of geographical enterprise in Spenser's time, when islands *did* move around uncertainly on maps and charts. Spenser displays a perhaps instinctive understanding that travel by water is characterized (at least in sixteenth-century terms) by both physical and epistemological fluidity.

Whether it is a river or a sea, the Idle Lake is a central feature of Book 2, offering a natural division between the earlier and later cantos. The division is a significant aspect of Guyon's journey; upon crossing the Idle Lake, he faces very different dangers and demands than what he faced on the other side, the side that reaches conclusion in canto 6. Different in what sense? Often the major *loci* in Book 2 mirror one another: Medina's castle and the House of Alma, Phaedria's island and the Bower of Bliss. The duels in cantos 5 and 8 involve the same villains, Pyrocles and Cymocles. Yet in the middle of Book 2 there is an important shift in the context of the action, one more subtle than the "sharp break dividing Book II into two sections which seem to have little in common" identified by Harry Berger many years ago.[9] In terms of Guyon's relations with the world around him, it is possible to distinguish between the *social* and the *solitary* cantos in Book 2.

A look at the cantos in sequence will lend more meaning to this distinction. I want to convey a sense of the multiplicity of the first half of Book 2.[10] The movement of the narrative is more cyclical than linear: characters appear, intersect, recede, reappear, locked in a martial (and occasionally venereal) *danse ronde*. Sometimes these characters—such as Phedon at 4.17–33—recite stories as complex as the actions in which they themselves are presently involved. The effect is of a world of tumultuous relationships with Guyon more or less at its center. Whether or not the reader treats this part of the book as a classic example of *psychomachia*, he or she has ample opportunity to compare and contrast Guyon with figures that surround him. At the same time, Spenser steadily forces Guyon out of the social field in which he initially operates.

In the first canto, Guyon (accompanied by the Palmer) meets and is duped by Archimago and Duessa; confronts and nearly fights with Redcross; and encounters Amavia, Ruddymane, and the corpse of Mordant. In canto 2 he dines with Medina, Elissa, and Perissa along with the suitors Huddibras and Sans-Loy, having already prevented these last two from duelling to the death. Canto 3 offers a comic *divertissement* involving

Braggadocchio, Trompart, Archimago, and Belphoebe. In canto 4 Guyon struggles with Furor and Occasion in order to rescue Phedon; Phedon then recounts at length the tale of himself, Philemon, Pryene, and Claribell. Finally, Atin appears as harbinger to Pyrocles, threatening Guyon with a dart. Canto 5 opens with Guyon fighting and defeating Pyrocles; he unbinds Occasion only to have her incite Furor to attack Pyrocles once again. Atin, having witnessed Pyrocles' initial defeat, goes to summon Cymocles from the Bower of Bliss. Canto 6 finds Phaedria ferrying Cymocles to her island, as she does later for Guyon. She narrowly prevents Guyon and Cymocles from grappling and then ferries Guyon over the remainder of the lake. Guyon has a brief scrape with Atin, who later discovers Pyrocles in agony after his battle with Furor. Archimago emerges to cure Pyrocles of his condition.

By the time Guyon reaches the other side of the Idle Lake, he has lost both his horse (2.11) and "his trusty guide," the Palmer (7.2, 6.19–20). I will later address the larger significance of these two losses. In canto 7, at any rate, Guyon faces the perils of Mammon's cave in isolation from both accustomed friends and accustomed enemies. The characters he meets in the cave are unlike the characters in the previous cantos in the sense that their allegorical and mythological significance is more important than the way they behave toward Guyon and each other. Figures like Disdain, Philotime, and Tantalus are static, pictorial. Pyrocles, Cymocles, and the Palmer (all far more "active" characters) return in canto 8, and Arthur is first introduced in Book 2, but Guyon remains unconscious until the end of the canto, when Arthur has already dispensed with the two "sonnes of *Acrates*." In effect, Guyon is "somewhere else" during most of the action. Arthur, furthermore, does not function like the characters in cantos 1–6. Arthur's appearance is heralded by an angel (8.8) and signals not only Guyon's spiritual rescue (echoing Arthur's rescue of Redcross from Orgoglio's dungeon in canto 8 of Book 1) but also a general heightening of tone. While it is true that the Palmer accompanies Guyon for the rest of his journey, this fact does not diminish the quality of solitude that pervades the last cantos. It becomes difficult, at times impossible, to overlook the allegorical aspects of the narrative from canto 7 onward—something it *is* possible to do in many of the earlier cantos. The House of Alma reads differently from Medina's castle because, for one thing, Guyon becomes an observer instead of an actor. Indeed, Guyon spends the whole of canto 10 reading—one of the most solitary and least active endeavors, particularly for an Arthurian knight.

This is not to suggest that Spenser reduces Guyon to a passive interpreter of an allegorical tableaux. He is still free to act, but this action comes outside the context established in the first half of Book 2. The exchange becomes one-sided; in canto 12, for instance, Guyon acts physically upon the Bower far more than the Bower acts physically upon him. If canto 11 offers a kind of counterpoint to canto 12—Spenser indicates that Arthur's battle with Maleger is concurrent with Guyon's voyage (11.5)—it also presents a similar phenomenon. Ferocious as the battle is, it remains curiously abstract in design, largely because there is no social standard of comparison between Arthur and his foe. While Arthur's victory over Pyrocles and Cymocles has an overt allegorical meaning—the triumph of the complete man over the sins of wrath and lust—it also presents itself at a literal level as the triumph of a superior knight over inferior competitors. Arthur's defeat of Maleger is more difficult to interpret literally; it reads more satisfactorily as a representation of a generalized conflict (health vs. disease, man vs. the elements, the soul vs. sin) than as an instance of hand-to-hand combat. Though Arthur (significantly, given the allegory) needs the help of his squire to stave off defeat (11.30–31), the prevailing impression is of Arthur struggling alone in an unreal and depopulated landscape—depopulated, at least, of any figures save allegorical ones.[11]

In fact, with the possible exceptions of Acrasia and Verdant, the characters in the late cantos function mainly as personifications and barely as persons. Alma, Mammon, the Boatman, and Genius are decidedly unlike Amavia, Cymocles, Phaedria, and Braggadocchio (not that the early cantos lack their share of straightforwardly allegorical characters such as Medina, Furor, Occasion, and Atin). The characters in the early cantos appear to inhabit the same plane of being as Guyon does, while those later on operate at some remove, as if separated from Guyon by a screen of "significance" (or, perhaps more appropriately, signification).

This account ought to confer additional sense on the physical layout of Book 2 as an element in Spenser's effort to present Guyon as a temperate knight in an intemperate world. But intemperance appears in both domestic and exotic forms, which one might read as Old World and New World. Guyon's journey across the Idle Lake separates him from a world of social relationships as well as comparative relationships. On the other side of the lake he is close to being alone in the wilderness. Where much of the moral energy in cantos 1–6 is generated by the direct friction between Guyon and the characters surrounding him, the energy in cantos 7–12 comes from a more nebulous conflict with what I call, for lack of a better phrase, "ambient otherness." Spenser concentrates his attention on Guyon as an isolated being, allowing the initial criteria of judgment concerning his character to drop away gradually until only this otherness remains, localized at last in the Bower of Bliss. Guyon's voyage to the Bower is the final separation from the social context established at the beginning of Book 2. Guyon arrives there with only the Palmer and his own resources for guidance, but the episode is anything but conclusive (as the critical debate over it amply demonstrates). The problem of Guyon's behavior in the Bower has dominated modern criticism of Book 2; what has often been overlooked is that the problem occupying Spenser's readers occupies

Spenser as well. Whether or not he solves it adequately—and whether it can truly be solved—remains to be seen. At this point, though, I want to look at the way in which the peripheral characters in Book 2 set the stage for Guyon's activities in solitude.

The social cantos of Book 2 are occupied with knights who fail dismally to approach an acceptable standard of knightly behavior: Huddibras, Sans-Loy, Braggadocchio, Pyrocles, and Cymocles. Each allegorically represents some form of wavering from the temperate mean represented by Guyon; at the same time, these versions of intemperance are historically charged. The knights above belong within a larger category of bad knighthood, which includes all the qualities implied in the allegorical names that Spenser applies to them: foolhardiness, lawlessness, vainglory, irascibility, concupiscence.

Such qualities are demonstrated in Book 2 as deeds of brutal and irrational violence. On the slightest pretext, Huddibras and Sans-Loy turn from fighting each other to assaulting Guyon. Spenser describes them not in human but in bestial terms, as "a Beare and Tygre" (2.22.5). Interestingly enough, Spenser uses a nautical simile to describe Guyon's riposte (2.24); the "lybicke Ocean" (2.22.6) has become an ocean indeed. Later a mounted Pyrocles attacks Guyon while Guyon is on foot; when Guyon, defending himself, accidentally slays Pyrocles' horse, Pyrocles accuses him absurdly of a cowardly act (5.5). Likewise, Cymocles descends upon Guyon merely because Guyon happens to be standing next to Phaedria on the shore of her island (6.28). In canto 8, Pyrocles and Cymocles attempt to desecrate Guyon's apparently lifeless body, offering the most specious reasons for their act:

> Loe where he now inglorious doth lye,
> To proue he liued ill, that did thus foully dye.
>
> The worth of all men by their end esteeme,
> And then due praise, or due reproch them yield;
> Bad therefore I him deeme, that thus lies dead on field.
> (8.12.8–9, 8.14.7–9)

The reason for this attempt is less a desire for vengeance than a desire to spoil Guyon of his valuable armor—an honorable trophy in battle, but here reduced by the brothers to the reward for rifling a corpse. As Pyrocles says, "why should a dead dog be deckt in armour bright?" (8.15.9). The brothers meet their end as a result of yet another rash act of violence, Pyrocles' attack on the unprepared Arthur. Before dispatching them, Arthur throws out an unequivocal indictment:

> False traitour miscreant, thou broken hast
> The law of armes, to strike foe undefide.

> But thou thy treasons fruit, I hope, shalt taste
> Right sowre, and feel the law, the which thou hast defast.
> (8.31.6–9)

Under this condemnation, Pyrocles and Cymocles are not knights at all, but rather outlaws to knighthood; at best they are cruel, rapacious, unreflecting hotheads; at worst, virtual graverobbers.

In the late Elizabethan period, the "dossier" of categorically bad knighthood bore the seal of the Habsburgs in its Spanish branch. To England and the rest of Protestant Europe, Philip II, like Huddibras, was "not so good of deedes, as great of name" and "More huge in strength than wise in workes" (2.17.3, 6); like Sans-Loy, "Ne ought he car'd, whom he endamaged / By tortious wrong, or whom bereau'd of right" (2.18.7–8); like Cymocles, he was "of rare redoubted might, / Famous throughout the world for warlike prayse, / And glorious spoiles, purchast in perilous fight," but ruthless with enemies, "Whose carkases, for terror of his name, / Of fowles and beastes he made the piteous prayes" (5.26.1–3, 6–7). Whether rightly or wrongly, Philip's Spain was transformed into the exemplar of intemperance on a global scale.

In 1580, for instance, William of Orange concluded his *Apologia* by describing "the banishments, the thefts of goods, the imprisonments, the torments suffered, the horrible and miserable deaths, through which this bloody people, more cruel than Phalaris, Busiris, Nero, Domitian, and all similar tyrants, have persecuted the poor [of the Netherlands]."[12] This is resounding hyperbole; Spain does not merely approximate, it surpasses, the most notoriously despotic regimes of antiquity. Richard Hakluyt the Younger's manuscript treatise of 1584, usually entitled the *Discourse of Western Planting,* does little to soften this view: "yea such and so passing strange and exceeding all humanity and moderation have they been that the very rehearsal of them drove diverse of the cruel Spanish which had not been in the west Indies, into a kind of ecstasy and maze."[13] Roughly a decade later, Thomas Nashe, satirizing the pride of various nations in *Pierce Penilesse his Supplication to the Divell,* turns his attention to Spain with similar scorn of its excesses:

> properly Pride is the disease of the Spaniard, who is born a Braggart in his mothers womb: for, if he be but 17 years old, and hath come to the place where a Field was fought (though half a year before), he then talks like one of the Giants that made war against Heaven, and stands upon his honor as if he were one of *Augustus* Soldiers, of whom he first instituted the order of *Heralds:* and let a man sooth him in this vein of killcow vanity, you may command his heart out of his belly, to make you a rasher on the coals, if you will next your heart.[14]

The concluding clause is more than a sardonic tag; it implies that a typical Spaniard is capable of the grimmest sort of behavior.

There are instances, too, in which the excesses of the Spanish are transmuted into the virtues of the English. The 1589 *A Summarie and True Discourse of Sir Frances Drakes West Indian Voyage,* an account by various hands of Drake's 1585–86 expedition, includes a possibly apocryphal anecdote about the furnishings of the viceroy's palace at Santo Domingo:

> I may not omit to let the world know one very notable mark and token of the unsatiable ambition of the Spanish King and his nation, which was found in the kings house, whereof the chief Governor of that City and country is appointed always to lodge, which was this. In coming to the hall or other rooms of this house, you must first ascend up by a fair large pair of stairs, at the head of which stairs is a handsome spacious place to walk in somewhat like unto a gallery, wherein upon one of the walls, right over against you as you enter the said place, so as your eye can not escape the sight of it, there is described & painted in a very large Scutcheon, the arms of the king of Spain & in the lower part of the said Scutcheon, there is likewise described a globe, containing in it the whole circuit of the sea and the earth, whereupon is a horse standing on his hinder part within the globe, and the other fore part without the globe, lifted up as it were to leap, with a scroll painted in his mouth, wherein was written these words in Latin NON SVFFICIT ORBIS: which is as much to say, as the world sufficeth not, whereof the meaning was required to be known of some of those of the better sort, that came in commission to treat upon the ransom of the town, who would shake their heads, and turn aside their countenance in some smiling sort, without answering any thing as greatly ashamed thereof. For by some of our company it was told them, that if the Queen of England would resolutely prosecute the wars against the king of Spain, he should be forced to lay aside that proud and unreasonable reaching vein of his, for he should find more than enough to do, to keep that which he had already, as by the present example of their lost town they might for a beginning perceive well enough.[15]

Drake is once again the instrumental figure in this cleverly designed tale, in which a world that "sufficeth not" evolves into one that is "more than enough" for the Spaniards. The story rests on a familiar rationale for the depredations of the English privateers: in sacking Santo Domingo, Drake's men are relieving the Spaniards of excess (whether this excess takes the form of treasure or ambition) in order to bring them back toward the mean; they are, in short, agents of temperance, improving the moral health of their enemies by plundering them.

The group of intemperate knights in Book 2 of *The Faerie Queene* serves for Guyon as, more broadly, Spain serves for England in accounts like the one above: that is, as a foil against which the intentions and deeds of the favored party will be certain to shine. To borrow some familiar terms from dialectical philosophy, these bad knights together form an antithesis that opposes the "thesis" of Guyon. He is the anti-conquistador to these conquistadors.

The synthesis that emerges from this opposition is, however, problematical. Of the three sword fights that befall Guyon in the first six cantos,[16] the first and last are interrupted by Medina and Phaedria respectively (2.27, 6.32), and the second, Guyon's victory over Pyrocles, is a demonstration more of force than of superior virtue. When Pyrocles' messenger Atin arrives to claim the ground surrendered by Occasion and Furor, Guyon responds, "Varlet, this place most dew to me I deeme, / Yielded by him, that held it forcibly" (4.40.1–2). Guyon's own use of force has entitled him to the spot. Throughout his subsequent duel with Pyrocles (5.3–12) there is no indication that Guyon possesses any sort of superiority other than of strength and endurance. Guyon literally presses Pyrocles into submission: "through great constraint / He maid him stoup perforce vnto his knee, / And do vnwilling worship to the Saint, / That on his shield depainted he did see" (5.11.5–8). Pyrocles is compelled to worship the Fairy Queen; Guyon makes no effort to persuade him that such worship is justified in any way. In fact, Guyon speaks to him with evident condescension, as a stronger contestant to a weaker: "Be nought agrieu'd, / Sir knight, that thus ye now subdewed arre: / Was neuer man, who most conquestes achieu'd / But sometimes had the worse, and lost by warre, / Yet shortly gaynd, that losse exceeded farre" (5.15.1–5). Pyrocles is a conqueror conquered. The gain that Guyon proposes to him is simply that he become more temperate—advice more insulting than consoling to the vanquished.

Now force of arms is in itself a neutral value, equally applicable to good or bad ends. The problem is that Guyon's metaphysical, as opposed to physical, excellence presents itself either as a given—assumed to be in Guyon's possession from the beginning of Book 2—or by way of comparison with the obvious moral, spiritual, and political inferiority of figures like Pyrocles and Cymocles. In neither case does Guyon's excellence emerge as proven. Both aspects of the problem surface in Spenser's postscript to the misadventures of Braggadocchio and Trompart:

> In braue pursuit of honorable deed,
> There is I know not what great difference
> Betweene the vulgar and the noble seed,

> Which vnto things of valorous pretence
> Seemes to be borne by natiue influence;
> As feates of armes, and loue to entertaine,
> But chiefly skill to ride, seemes a science
> Proper to gentle bloud; some other faine
> To menage steeds, as did this vaunter; but in vaine.
> (4.1)

This is a fairly subtle piece of question-begging. What is the nature of this "great difference" between "the vulgar and the noble seed"?[17] The difference, Spenser answers, is that the noble seed is naturally noble. What are the outward signs of this nobility? Not holy, chaste, or even temperate behavior, but "feates of armes," "loue," and "chiefly skill to ride." Spenser's demonstration of this latter skill takes a negative form, from the "vulgar" example of Braggadocchio. The positive demonstration never occurs, since Guyon remains unhorsed for the rest of Book 2. There is a circularity to this stanza; it fails to make much of a case for nobility or, more importantly, for Guyon as a noble figure. What is the source of Guyon's goodness? What values does he represent? Spenser allows these questions to remain far into Book 2.

Pyrocles and Cymocles are identified as "Sarazin" and "Paynim" (8.49.1, 8.50.1) and make oaths on the names of "*Termagaunt*" and "*Mahoune*" (8.30.4, 8.33.3).[18] At the most basic level they are stereotypical Muslim marauders, the familiar villains of innumerable medieval and Renaissance narratives, including those that most clearly influenced *The Faerie Queene*'s grand design, Ariosto's *Orlando Furioso* and Tasso's *Gerusalemme Liberata*. However, there was a pronounced tendency in Spenser's day to confuse the Spaniards with their former oppressors. The introduction to *The Spanish Colonie* displays one type of this confusion: "Thou shalt (friendly reader) in this discourse behold so many millions of men put to death, as hardly there have been so many Spaniards since their first fathers the Goths inhabited their Countries, either since their second progenitors the *Saracens* expelled and murdered the most part of the Goths."[19] There are no authentic Spaniards, only the offspring of two different waves of barbarians. Spenser follows a similar tack in *A View of the Present State of Ireland:*

> after all these the Moors and Barbarians, breaking over out of Africa, did finally possess all Spain, or the most part thereof, and did tread, under their heathenish feet, whatever little they found yet there standing. The which, though afterward they were beaten out by Ferdinando of Aragon and Elizabeth his wife, yet they were not so cleansed, but that through the marriages which they had made, and mixture with the people of the land, during their long continuance there, they had left no pure drop

> of Spanish blood, no more than of Roman or Scythian. So that of all the nations under heaven (I suppose) the Spaniard is the most mingled, and most uncertain.[20]

The Spaniards have in effect become the modern Saracens. This notion receives a further airing in Book 5 of *The Faerie Queene,* where in canto 8 Spenser allegorizes the defeat of the Armada through Arthur's battle with the Souldan (5.8.28ff.).[21] The Souldan bears comparison with Pyrocles and Cymocles; like them he melds sacrilege and savagery in his actions, "Swearing, and banning most blasphemously" as he prepares for battle in a chariot drawn by carnivorous horses, "which he had fed / With flesh of men, whom through fell tyranny / He slaughtred had, and ere they were halfe ded, / Their bodies to his beasts for prouender did spred" (5.8.28.2, 6–9)—a description that savors faintly of *la leyenda negra.*

Further connections can be drawn from the examples of Pyrocles and Braggadocchio, in both cases arising from the relationship between the character and a more primitive counterpart: Furor and Trompart, respectively. Furor is not only a personification of his name; he is also, quite obviously, a savage, full of "beastly brutish rage," battling Guyon with "rude assault and rugged handeling," pummeling him with "clownish fistes" (2.4.6.7, 4.8.1, 4.9.2). He has no "gouernance" (4.7.2) and his appearance is decidedly wild, with "long lockes, colour'd like copper-wire" and a "tawny beard" (4.15.8, 9). His closest kinship in Book 2 is with Maleger and his followers in canto 11, being "not such a foe, / As steele can wound, or strength can ouerthroe" (4.10.4–5). Though his complexion reflects his sanguinary temperament and could be construed as "Celtic," Furor also evokes some of the common stereotypes of the New World native—unreasoning, inscrutable, and incapable of surrender. Martyr, narrating Columbus's capture of several cannibals, portrays them in such a light: "When they were brought into the Admirals ship, they did no more put off heir fierceness and cruel countenances, then do the Lions of *Lybia* when they perceive them selves to be bound in chains. There is no man able to behold them, but he shall feel his bowels grate with a certain horror, nature hath endowed them with so terrible menacing, and cruel aspect."[22] Furor, like these cannibals, must be restrained with chains—a "hundred yron" ones are barely enough (4.15.1).

Pyrocles is less fortunate than Guyon in his encounter with Furor. I want to advance the possibility that Pyrocles is not only figuratively but literally poisoned by Furor. His agonized account of his symptoms certainly describes the effects of poison: "His [i.e., Furor's] deadly wounds within my liuers swell / And his whot fire burnes in mine entrails bright, / Kindled through his infernall brond of spight" (6.50.3–5). He is already in desperate straits when Guyon departs the scene (5.23–24), and when Atin discovers him at the end of canto 6 he is suffering terrible torments:

> I burne, I burne, I burne, then loude he cryde,
> O how I burne with implacable fire,
> Yet nought can quench my inly flaming syde,
> Nor sea of licour cold, nor lake of mire,
> Nothing but death can doe me to respire.
> (6.44.1–5)

Allegorically, Pyrocles illustrates the effects of irascibility; the waters of the Idle Lake (and the vice of sloth symbolized by those waters) can do nothing to mitigate the agony that wrath makes him suffer. Physically, Pyrocles has been wounded by Furor's "flaming fyre brond" (5.22.6), the tokens of which are already implied in the name of the victim. Indeed, *The Faerie Queene* is full of characters who suffer rankling internal wounds; what is unusual in Pyrocles' case is the way in which he tries to cure himself, having "deepe him selfe beducked" in the Idle Lake (6.42.3).

This rhymes interestingly with contemporary references to a remedy for the one indigenous weapon truly feared by the conquerors of the Americas: the poison with which the "Cannibals" tipped their arrows. One of Richard Eden's excerpts from Oviédo's *Historia* comes under the heading "Of venomous apples wherewith they poison their arrows":

> In so much that if it may be spoken of any fruit yet growing on the earth, I would say that this was the unhappy fruit whereof our first parents Adam and Eve tasted, whereby they lost their felicity and procured death to them and their posterity. Of these fruits . . . the Cannibals which are the chief archers among the Indians, are accustomed to poison their arrows wherewith they kill all that they wound. . . . whereas the Christians which serve your majesty in these parts, suppose that there is no remedy so profitable for such as are wounded with these arrows, as is the water of the sea if the wound be much washed therewith, by which means some have escaped although but few, yet to say the truth, albeit the water of the sea have a certain caustic quality against poison, it is not a sufficient remedy in this case: nor yet to this day have the Christians perceived that of fifty that have been wounded, three have recovered.[23]

It is worth stressing not the precise connection between Pyrocles' wound and the poisoned arrows, but the idea that Spenser addresses the very material consequences of an encounter between an intemperate knight and an equally intemperate "salvage man," Furor—an encounter the structure of which nicely echoes a notorious (and frequently reported) type of confrontation between conquistador and native in the New World.

The Braggadocchio episode in canto 3 demonstrates another aspect of the relationship between the intemperate knight and the savage.

Spenser's description of Braggadocchio employs terms from the mining trade, both anticipating the subject matter of canto 7 and suggesting that Braggadocchio's vainglory is a form of greed, not unlike the greed for precious metals: "Ne thought of honour euer did *assay* / His *baser* brest, but in his kestrell kind / A pleasing *vaine* of glory vaine did find" (3.4.3–5; emphasis mine). Braggadocchio's elevation to "knighthood" results less from his audacious self-presentation than from his theft of Guyon's horse; it is after this theft that he becomes "puffed vp with smoke of vanitie" (3.5.3). His victory over the "seely man" Trompart is largely because of the horse: upon seeing Braggadocchio "ride so rancke," Trompart falls "flat to grounde for feare" (3.6.7, 8). Braggadocchio's speech to his prostrate victim echoes Pyrocles' at 8.10.9 to the equally prostrate (and still horseless) Guyon: "Why liuest thou, dead dog, a lenger daye" (3.7.6). As a token of Trompart's submission, Braggadocchio orders him to "kisse my stirrup" (3.8.6). The association between Guyon's horse and Braggadocchio's comic ascendancy is intriguing in light of the significance of horses to those "knights" who had crossed the Atlantic in search of their fortunes and of the prestige that came with conquest. The sixteenth-century conquistadors depended heavily on cavalry to suppress native populations in the Americas; the Indians had never seen horses before, were terrified of them, and usually failed to mount an effective defense against them. Both Gómara's and Bernal Díaz del Castillo's reports concerning the battle at Centla, one of the first battles of the Spanish invasion of Mexico in 1519, state that the Mayas with whom Cortés fought had fled before a small group of mounted soldiers, thinking that the horses and riders were single animals, like centaurs.[24] In the account of John Hawkins's 1564 voyage to the Indies printed in the 1589 edition of Hakluyt's *Principall Navigations,* the writer observes that on an open plain "one horseman may overrun 100. of" the warlike Caribs; he does add that the Caribs have become "politic warriors," having learned how to place woodpiles and pikes across the level ground to "mischief" the Spanish horses.[25]

Trompart shows himself to be politic in a somewhat different sense when faced with a "knight" lording it over him on horseback. He quickly learns that Braggadocchio's power resides in the trappings, not in the man himself:

> [W]hen he felt the folly of his Lord,
> In his owne kind he gan him selfe vnfold:
> For he was wylie witted, and growne old
> In cunning sleights and practick knauery.
> (3.9.3–6)

The phrase "In his owne kind" and the observation that Trompart is "growne old" in the arts of deception and treachery both suggest that Spenser is trying to draw a generic, even racial, distinction here; these qualities are characteristic not only of Trompart but also of a larger group to which he belongs. The passage also suggests that the slave is quite capable of dominating the master, if the master is as inadequate to the task as Braggadocchio happens to be.

The other crucial character in canto 3 is Belphoebe, whose appearance clearly demonstrates Spenser's method of displaying an exemplary thesis against an antithetical background. Belphoebe is an imposing and glorious figure, a vision of Old World nobility transmuting a savage wilderness into a civil Arcadia. As obvious as Belphoebe's qualities are, they are enhanced and heightened in contrast to an impostor like Braggadocchio. He is a self-regarding foil like Bottom or Sir Andrew Aguecheek in Shakespeare's comedies, setting off the virtues of his betters to comic advantage. Spenser's treatment of him is not very different from a popular burlesque of the Spaniard noted by Hakluyt in the *Discourse of Western Planting:*

> The Italians . . . do many ways show the utter mislike of their [the Spaniards'] satanical arrogancy and insolences, and in all their plays and comedies bring in the Spanish soldier as a ravisher of virgins and wives, and as the boasting Thraso and *miles gloriosus;* noting to the world their insupportable luxuriousness, excessive pride and shameful vainglory.[26]

Braggadocchio is not only a *miles gloriosus* but also (in intent at least) a ravisher of virgins, attempting unsuccessfully to rape Belphoebe: "the foolish man . . . with her wondrous beautie rauisht quight / Gan burne in filthy lust, and leaping light, / Thought in his bastard armes her to embrace" (3.42.2, 4–6). Trompart, on the other hand, shows the tendency of the savage to attribute divinity to noble mortals. "O Goddesse" initiates his address to Belphoebe, and he asks her "which of the Gods I shall thee name, / That vnto thee due worship I may rightly frame" (3.33.2, 8–9); later he speculates, "who can tell . . . / But that she is some powre celestiall?" (3.44.3–4). This attitude repeats, at a more sophisticated level, that of the "salvage nation" toward Una in Book 1: "The woodborne people fall before her flat, / And worship her as Goddesse of the wood" (1.6.16.1–2). In this attribution of godhood, though, Trompart is more "politike" than is Braggadocchio in his lust, for Spenser declares that Belphoebe seems "borne of heauenly birth" (3.21.9).

Spenser's depiction of Belphoebe is often viewed as one of his most obvious tributes to Elizabeth. I want to consider briefly what other purposes are served by her introduction into the poem. First, she is ideally suited to the subjection of "salvage" men and nations. Where Braggadocchio and Trompart nearly panic upon entering the forest—"Each trembling leafe, and whistling wind they heare, / As ghastly bug their haire on end does reare" (3.20.4–5)—Belphoebe is entirely at home in the wilderness. Spenser compares her to the "Queene / Of *Amazons*" (3.31.5–6), which can be understood as either queen among the Amazons or queen over the Amazons. In either case, the phrase resonates with Spenser's mention of the river Amazon in the proem, and also with the accumulating lore of the New World, which transposed the Amazons of classical mythology to various places in the Americas. Belphoebe's homily on honor is a rallying cry for the sort of gentleman-adventurer-discoverer whom the English expected would eventually wrest from Spain the glories and riches of the New World:

> Who so in pompe of proud estate . . .
> Doth swim, and bathes himself in courtly blis,
> Does waste his dayes in darke obscuritee . . .
> Where ease abounds, yt's eath to do amis . . .
> Abroad in armes, at home in studious kind
> Who seekes with painfull toile, shall honor soonest find.
>
> In woods, in waues, in warres she wonts to dwell,
> And will be found with perill and with paine . . .
> Before her gate high God did Sweat ordaine,
> And wakefull watches euer to abide:
> But easie is the way, and passage plaine
> To pleasures pallace. . . .
> (3.40.1–3, 5, 8–9, 3.41.1–2, 5–8)[27]

This is similar in tone to a passage in Nicholas's introduction to his translation of Zárate. Nicholas's foreword is essentially propaganda for English colonialism; speaking of Sir Francis Drake, he says,

> His painful travail, and marvelous Navigation, was not obtained with white hands, perfumed gloves, dainty fare, or soft lodging: no, no: *Honor* is not gotten with pleasures, & quiet minds. For the sweet Roses groweth among Thorns: yet the ignorant will judge, that perpetual Fame and heavenly Felicity, is a thing to be gotten with facility and ease. But the poor Sailor should sit as Judge, I am sure that he would say, how extreme hunger, thirst, hard lodging upon Hatches, fouled garments, blustering storms of wind, with hail, Snow, bitter cold, Thunder Lightning, and continual peril of Life, leadeth the high pathway to *the Court of Eternal Fame*.[28]

Belphoebe's speech can stand alone as a commentary on the pursuit of both personal and national honor; it serves as well to make Braggadocchio's earlier speech on his own pursuit of honor, in which he claims he has "many battles fought . . . / Throughout the world . . . / Endeauouring my dreadded name to raise / Aboue the Moon" (3.38.5–8), seem far more ludicrous.

This leads to a second, equally important purpose served by Belphoebe's presence in the episode. The relationship between Belphoebe and Braggadocchio is retroactive: not only does Braggadocchio make Belphoebe look even better, but Belphoebe also makes Braggadocchio look even worse. Belphoebe's role takes on further meaning in the context of the intention, expressed by Hakluyt among others, to use colonial enterprise as a way of exposing the Spaniards' true character:

> I may therefore conclude this matter with comparing the Spaniards unto a drum or an empty vessel, which when it is smitten upon yieldeth a great and terrible sound and that afar off, but come near and look into them, there is nothing in them, or rather like unto the ass which wrapped himself in a lions skin and marched far off to strike terror into the hearts of other beasts, but when the fox drew near he perceived the long ears and made him a jest unto all the beasts of the forest. In like manner we (upon peril of my life) shall make the Spaniard ridiculous to all Europe, if with piercing eyes we see into his contemptible weakness in the west Indies, and with true style paint him out *ad vivum* unto the world in his faint colors.[29]

Spenser uses Belphoebe to paint, as it were, with "true style" the "faint colors" of Braggadocchio—and the indictment of Braggadocchio can be extended to all of the other intemperate knights of Book 2.

CHAPTER 4

Hunger of Gold: Guyon at the Cave of Mammon

Quid non mortalia pectora cogis, auri sacra fames!

Aeneid

By following Guyon into and out of the Cave of Mammon, I want to consider more intently the paradox that looms once Guyon's character is understood in the historical terms that I have previously outlined. The paradox is by now a familiar one: Guyon's distinction from the model offered by the conquistador rests at the same time upon a basic consonance with that model. In a formal sense, Guyon does most of the things that conquistadors do, but he does them with an "English" difference. His national origin is supposed to place him above the fray, as it were. The general effect is of Guyon incarnated as a Spaniard, but a Spaniard immaculately conceived. This sort of conception, of course, raises more questions than it answers. The following is an attempt to read more closely on the question of Guyon's strange historical and figurative "between-ness" at the site where, outside of the Bower of Bliss, the question becomes most pointed: where what the rest of Europe saw as the central object of Spanish colonization in the New World—the acquisition of gold—is also the main element of Spenser's allegory and Guyon's ordeal.

As Guyon emerges on the far side of the Idle Lake, Spenser once again draws a nautical analogy to his protagonist's condition, describing him as a "Pilot" who, because the skies are obscured, must depend on the ship's own resources, the "card and compas," to navigate safely (7.1.1, 6). Likewise, by contemplating his "owne vertues, and prayse-worthy deedes," Guyon manages to guide himself after a fashion "through wide wastfull ground" and "desert wildernesse" (7.2.5, 8, 9).[1] When the image in the first stanza, the self-contained ship separated from the wide skies overhead, is converted into the most general terms, the effect is of a small, interior world being distinguished from a large, expansive one. This distinction can be interpreted, and indeed has been interpreted, in many ways; the Cave of Mammon is one of the most thoroughly examined topoi in *The Faerie Queene* as a whole. Most commentators agree that the episode turns upon the conflict between Guyon the "inner man" and an outer world full of material temptations. Here the critical paths diverge; they meet again at the point where Guyon faints after leaving the Cave, but the meeting of the critics at the crossroads has not been a particularly friendly one.[2]

At the simplest level—and this is often a better point of departure than one might think—the episode of the Cave of Mammon depicts the evils attendant upon the lust for wealth, specifically the lust for gold.[3] Spenser composed it at a time when the main source and conduit of this kind of wealth for all Europe was the *Casa de Contratacion* at Seville. Gold from Peru and silver from Bolivia (from the great mine at Potosí, discovered in 1545), entering new markets in—to borrow Mammon's expression (7.8.8)—an "ample flood" of Spanish currency, had begun to alter irrevocably the economies of the Old World. J. H. Elliott notes the effect of Spain's vast treasury on her neighbors: "This inescapable fact impressed, and indeed overimpressed, contemporaries." He goes on to speak of "the obsessive preoccupation of the age with the silver of the Indies as the key to Spanish power."[4] Thus for images of wealth and its evils, Spenser need not have turned to the literary remains of past, for present-day political and economic conditions provided an ample and vivid supply.

The image, for instance, of Mammon sitting on his horde acquires sharper definition when viewed as a parody of the process by which Spaniards acquired wealth in the New World:

> And round about him lay on euery side
> Great heapes of gold, that neuer could be spent:
> Of which some were rude owre, not purifide
> Of *Mulcibers* deuouring element;
> Some others were new driuen, and distent
> Into great Ingoes, and to wedges square;
> Some into round plates withouten moniment;
> But most were stampt, and in their metall bare
> The antique shapes of kings and kesars straunge and rare.
> (7.5)

This stanza provides a digest of the whole procedure of rendering gold into usable currency, a procedure that was almost the sole property of the Spaniards in the late sixteenth century.[5] Spenser alludes to the process quite accurately: Indies bullion ("rude owre") was taken to the crown's local assay offices where it was smelted ("new driuen, and distent") into more easily transportable shapes ("ingoes . . . wedges square . . . round plates withouten moniment"). After shipment to Seville, the metal was minted ("stampt") into *coronas* and *reals*.[6] The "antique shapes of kings and kesars" carry implications of imperial powers for which such images had always been the "currency."[7] Certainly in Spenser's time, Philip II, drawing his inevitable share from the New World treasures, came the closest of any European monarch to being a modern "kesar." The fact that the ingots are "new driuen" suggests that Spenser is weighting his images with a degree of contemporary relevance.[8]

The textual sources that might have contributed to this effect of relevance have already been indicated. Hakluyt's *Principall Navigations* appeared in 1589, most likely too late to have much influence on the contents of *The Faerie Queene*. Nicholas's translations of Zárate and Gómara had been available since the early 1580s; *The Spanish Colonie* dates from the same period. Probably the most accessible and important geographical work in English prior to Hakluyt's was Eden's 1555 collation of Martyr's *Decades* and excerpts from other writers. This stands as the only major collection of New World information for the early-to-middle Elizabethan period. Hakluyt depended heavily on Eden in compiling his own collection. Spenser might readily have encountered Willes's 1577 revision of Eden's book, if not the original edition.[9]

One of Eden's obvious purposes in translating and assembling his materials was to satisfy a newfound need among English readers to know more about Spain's power beyond its own borders. Another, however, was to meet an English appetite for reading about gold and other precious metals; there are references to gold or gold-bearing minerals on nearly every page. As if the New World accounts were not rich enough on this subject, Eden appended selections from Vannoccio Biringuccio's *Pirotechnia*, originally published in 1540 and the standard reference work on metallurgy for the sixteenth and seventeenth centuries.[10] A survey of the material that Eden chooses to include reveals a curious ambivalence about gold and the value that the English ought to place on it, an ambivalence originating in Biringuccio's own work. In his introductory remarks, Biringuccio describes mining as a kind of moral crusade: "He therefore that hath begun to dig a cave, let him determine to follow it, putting away th[']estimation of the baseness thereof, and not to fear the straitness of the way, but rather to apply all his possible diligence without remorse, hoping thereby no less to obtain honor and riches, then to avoid shame and infamy for omitting so profitable an enterprise."[11] This is an odd passage, its oddness enhanced by Eden's translation, with its use of a scripturally allusive phrase like "the straitness of the way." The first object of the search for gold is honor rather than riches, and the diction of the passage implies that this is literally sacred honor. This unreserved enthusiasm for mining is almost immediately qualified in Eden's next excerpt from Biringuccio, describing "the mine of gold and the quality thereof in particular":

> this metal is a body tractable and bright, of color like unto the sun: And hath in it inwardly such a natural attractive or alluring virtue, that being seen, it greatly disposeth the minds of men to desire it and esteem it as a thing most precious: although many there are which cry out against it and accuse it as the root and seed of most pestiferous and monstrous covetousness, and the cause of many other mischiefs. But whether it be the cause of more good or evil, we intend to let pass this disputation as a thing unprofitable.[12]

Biringuccio avoids any moral discomfort on the subject of gold by mooting the question. Eden admits to the same discomfort in his preface to the *Pirotechnia* selections, but he confronts it in a very different way, calling

> gold and silver the seeds of all mischiefs and angels of such a god, whom the antiquity (not without good consideration) painted blind, affirming also that of him gold and silver have received the property to blind the eyes of men. But sith it is now so that we shall be enforced to seek aid by that which was sometimes a mischief, it resteth to use the matter as do cunning physicians that can minister poison with other things in such sort qualifying the maliciousness thereof, that none shall thereby be intoxicate.[13]

This is interesting in two respects. First, the money-god shares Cupid's blindness, recalling the double meaning of "cupidity" as both greed and lust, and perhaps indicating a reason that the Cave of Mammon and the Bower of Bliss find their way into the same book of *The Faerie Queene*. Second, Eden confesses that gold is a poison but proposes to treat it as a kind of vaccine which, used properly, becomes regenerative rather than destructive.

England had already benefitted from Spanish "poison" thanks to Drake's Caribbean raids of the 1570s. Objectively speaking, these raids were acts of piracy; under the pressure of patriotism they became the procedures of a "cunning physician" in turning gold to a proper use—that is, a use that served larger ends than mere appetite for gold. In the dedication "To the Reader" that prefaces his translation of Gómara, Nicholas contrasts Grijalva, who reconnoitered the Yucatan peninsula and the coastline above it prior to Cortés's expedition, with the idealized figure of Cortés himself, who seems to represent for Nicholas an exemplary aberration from the Spanish norm—a crypto-Englishman, so to speak: "This *Grijalva* pretended not to conquer, nor yet to inhabit, but only to fill his hungry belly with gold and silver, for if he had pretended honor, then *Cortez* had not enjoyed the perpetual fame which now is his, although his corpse be clothed in clay."[14] Here Nicholas suggests that the desire "to conquer" and "to inhabit" contributes signally to the honor of a man of action. Gold becomes honorable when it serves as a means to greater acquisitions—of lands, kingdoms, peoples—rather than as an end in itself. Of course, Spain by and large used gold as just such a means, a fact of which the English were certainly aware. Even so, Spain's actions were filtered through the English sense of Spain's degeneracy, so that these actions were more likely to be figured in a man like Grijalva, who could not see past his own "hungry belly," than in a man like Cortés.

The notion of hunger leads toward the vexed question of Guyon's faint at the end of canto 7. Spenser indicates that Guyon faints "For want of food, and sleepe, which two upbeare, / Like mightie pillours, this fraile life of man, / That none the same enduren can" (7.65.3–5). An alternate explanation that has been proposed and dismissed for many years is that Guyon's faint is the simple result of hunger and fatigue; for, after all, "three dayes of men were full outwrought, / since he his hardie enterprize began" (7.65.6–7). This explanation has largely been abandoned because it is so reductive; there is obviously a great deal more to say about the meaning of Guyon's collapse as a feature of Spenser's text. While the evidence for lack of sleep is inconclusive, I believe that a case can be made for literal hunger as significant to the outcome of canto 7. Spenser's use of hunger here is historically poignant, offering a more compelling account of Guyon's faint than a notion of "mere" hunger can provide.

Throughout Eden's translation of the *Decades* there are references to and marginal glosses on the lack of "vittles" that afflicted many of the early expeditions to the New World. Martyr frequently makes a direct connection between hunger and the quest for gold; at times that quest seems to have famine as its inevitable consequence.[15] The paradox of wealth creating a dangerous sort of poverty emerges clearly when Martyr writes, "Our men . . . went forward laden with gold, but sore afflicted with hunger."[16] The affliction is not only genuine hunger but a metaphorical "hunger" for gold that prevents or frustrates men from fulfilling their natural bodily needs.

This perversion of hunger also leads to the perversion of the true and honorable ends of conquest. "For," as Martyr says, "the hunger of gold, did no less encourage our men to adventure these perils and labors than did the possessing of the lands."[17] Gold-hunger, combined with other vices of the human character, leads to fiascoes of the sort that befell Columbus's brother Bartholomew, the "Adelantado," when he erected a fort near Santo Domingo on Hispaniola. The story as Martyr recounts it has the terseness of a parable: "This he called the golden tower, because the laborers found gold in the earth and stone whereof they made the walls of the fortress. He consumed three months in making the instruments wherewith the gold should be gathered, washed, tried, and molten. Yet was he at this time by reason of want of victuals, enforced to leave all things imperfect, and to go seek for meat."[18] The hunger for gold leads beyond real hunger to the worst sort of degradation; Martyr's narrative of Diego de Nicuesa's disastrous 1509 expedition to Darien, on what is now the Caribbean-side border between Panama and Colombia, offers many vivid examples. Nicuesa's men had even lost their identity as soldiers: "their strengths were so weakened with hunger, that they were not able to keep war against naked men, or scarcely to bear their harness on their backs."[19] They were reduced to eating dogs, frogs, mud, and, in one instance, the rotten corpse of a native, "therewith assuaging their hunger as if they had been fed with pheasants."[20] The Spaniards unwittingly became the cannibals they so much despised.

Finally, Rodrigo de Colmenares rescued Nicuesa's company from these desperate straits, although it was almost too late: "he found *Nicuesa,* of all living men most unfortunate, in manner dried up with extreme hunger, filthy and horrible to behold, with only three score men in his company, left alive of seven hundred. They all seemed to him so miserable, that he no less lamented their case, than if he had found them dead." Despite his condition, Nicuesa was still in the grips of an obsession that left little room for gratitude: "after that he had fallen down to the ground and kissed the feet of *Colmenares* his saviour, he began to quarrel with him . . . reproving him and them all . . . for the gathering of gold: Affirming that none of them ought to have laid hand of any gold with out [his] advice."[21] This portrait of Nicuesa bears comparison with Spenser's description of Mammon, "An vncouth, saluage, and vncivile wight, / Of griesly hew, and fowle ill fauour'd sight" (7.3.4–5), who goes from being "in great affright" and "terrifyde" at his first sight of Guyon (7.6.1, 7) to treating him with "great disdaine" moments later: "I read thee rash, and heedlesse of thy selfe, / To trouble my still seate, and heapes of precious pelfe" (7.7.6, 8–9).

The most compelling expression of the equation of greed with hunger in Martyr's text occurs in an encounter between the troops of Vasco Nuñez de Balboa and the son of a friendly chieftain. Martyr has also given this episode the shape of a moral fable; it even carries overtones of Christ's expelling the moneychangers from the temple.

> Here as brabbling and contention arose among our men about the dividing of gold, this eldest son of King *Comogrus* being present, whom we praised for his wisdom, coming some what with an angry countenance toward him which held the balances, he struck them with his fist, and scattered all the gold that was therein, about the porch, sharply rebuking them with words in this effect. What is the matter you Christian men, that you so greatly esteem so little a portion of gold more than your own quietness, which nevertheless you intend to deface from these fair ouches and to melt the same into a rude mass. If

> your hunger of gold be so insatiable that only for the desire you have thereto, you disquiet so many nations, and you your selves also sustain so many calamities and incommodities, living like banished men out of your own country, I will show you a Region flowing with gold, where you may satisfy your ravening appetites. But you must attempt the thing with a greater power: For it standeth you in hand by force of arms to overcome kings of great puissance, and rigorous defenders of their dominions. . . . or ever you can come thither, you must pass over the mountains inhabited of the cruel Cannibals a fierce kind of men, devourers of mans flesh, living without laws, wandering, and without empire. For they also, being desirous of gold, have subdued them under their dominion which before inhabited the gold mines of the mountains, and use them like bondmen, using their labor in digging and working their gold in plates and sundry images.[22]

Here is an explicit connection between the Spaniards and the cannibals: "For they *also,* being desirous of gold." To be compared with lawless, stateless, wandering slavers is a striking indictment of the Spaniards. Their industriousness in gathering gold notwithstanding, the cannibals still lack the necessities of life. Comogrus's son explains the trade between his people and the cannibals. "These things do we receive of them for exchange of our things . . . and especially for victuals whereof they stand in great need by reason of the barrenness of the mountains."[23] He goes on to speak of other vices that afflict his own tribe; again there is an underlying analogy to the vices of the Spaniards: "Albeit that the greedy hunger for gold hath not yet vexed us naked men, yet do we destroy one an other by reason of ambition and desire to rule. Hereof springeth mortal hatred among us, and hereof cometh our destruction."[24] The tale has a powerful coda: the Spaniards ignore the moral lesson concealed in the son's speech; their reaction is instead one of visceral hunger. "After these words, this prudent young *Comogrus* held his peace. And our men moved with great hope and hunger of gold, began again to swallow down their spittle."[25]

These examples help to suggest that Guyon's hunger at the conclusion of canto 7 is significant less for what it is than for what it is not; it is not gold-hunger, which blurs so easily and terribly into real hunger for the Spaniards. Harry Berger and Barbara-Maria Bernhart have each observed that canto 7 describes a perversion of the edible and inedible, distilled in the notion that Mammon "feede[s] his eye" on his golden horde (7.4.8).[26] Neither reader has noted that this perversion is part of the fabric of events at the sixteenth-century colonial frontier. The Spanish explorers literally and repeatedly famished themselves in order to sate their hunger for precious metals. What distinguishes Guyon from his historical predecessors is that his hunger is not caused by greed. Yet Guyon's experience resembles the Spaniards' in many other respects. Spenser has designed canto 7 so that Guyon undergoes an ordeal that should leave him reduced like the pathetic figure of Nicuesa, but does not. If physically reduced, Guyon is metaphysically intact. It is a matter not so much of temptations refused as of an identity transcended.

From the beginning of the canto, Guyon is distinguished by what he lacks: he is not a *caballero,* having lost his horse; he is missing religious authority or at least validity as represented by the Palmer, who is left waiting on the other side of the Idle Lake. In losing these appurtenances, Guyon has lost two of the conventional symbols of sixteenth-century empire, one military and one ecclesiastical, symbols the Spaniards carried with them throughout their New World expeditions. Guyon plays the conquistador as unaccommodated man, stripped of the usual attributes of conquest.

He immediately confronts a figure with the most extreme pretensions to imperial grandeur; he speaks like a king whose power, as he proclaims in the last line of stanza 8, comes from the mines.

> God of the world and worldlings I me call,
> Great *Mammon,* greatest god below the skye,
> That of my plenty poure out vnto all,
> And vnto none my graces do enuye:
> Riches, renowne, and principality,
> Honours, estate, and all this worldes good,
> For which men swinck and sweat incessantly,
> Fro me do flow into an ample flood,
> And in the hollow earth haue their eternall brood.

While Mammon claims to be a god, his language is redolent of monarchy; he is a "god below the skye" who speaks of graces, renown, honors, and estate. In the late sixteenth century, the "king of the mines" was, in effect, Philip II; his New World assets offer an appropriate analogy to Spenser's description of the House of Richesse, which

> shewd of richesse such exceeding store,
> As eye of man did neuer see before;
> Ne euer could within one place be found,
> Though all the wealth, which is, or was of yore,
> Could gathered be through all the world around,
> And that aboue were added to that vnder ground.
> (7.31.4–9)

The likelihood that Guyon is entering no mere cave but an actual mine has been little remarked, perhaps because it seems so obvious.[27] Many features of the episode suggest that Spenser intends Mammon's Cave to be perceived as a mine. Mammon's heaps of treasure are described as "pretious hils" (7.6.3) and "mountaines" (7.9.2), conjuring up images of the settings in which mines such as Potosí had been discovered. Spenser's depiction of the house of Richesse at stanzas 28–29 tends to confirm these initial images:

> That houses forme within was rude and strong,
> Like an huge caue, hewne out of rocky clift,
> From whose rough vaut the ragged breaches hong
> Embost with massy gold of glorious gift,
> And with rich metall loaded euery rift. . . .
>
> Both roofe and floore, and wals were all of gold,
> But ouergrowne with dust and old decay,
> And in darkenesse, that none could behold
> The hew thereof. . . .
>
> (7.28.1–5, 29.1–4)

The cave is "hewne" out of the rock; it is man-made. The ore within has not yet been touched and is still "ouergrowne with dust." There is activity in the mine, however, for Arachne, spinning her web on the ceiling, is "Enwrapped in foule smoke and clouds more blacke then Iet" (7.28.9). The presence of smoke should not be surprising, for in another room Guyon encounters a large smelting operation: "Therein an hundred raunges weren pight, / And hundred fornaces all burning bright" (7.35.4–5). The depiction in stanzas 28–29 recalls a traditional organic notion of the origins of gold that Martyr repeats in the *Decades:* "They say that the root of the golden tree extendeth to the center of the earth and there taketh nourishment of increase. For the deeper that they dig, they find the trunks thereof to be so much the greater as far as they may follow it. . . . They have sometimes chanced upon whole caves sustained and borne up as it were by golden pillars. And this in the ways by which the branches ascend."[28] There is another echo of this notion in Spenser's account of Philotime's court: "Many great golden pillours did vpbeare / The massy roofe, and riches huge sustayne" (7.43.5–6).

The fact that the entrance to the House of Richesse is next to "the gate of Hell" actually speaks in favor of the identification of Mammon's Cave with a mine. The river Cocytus, which Guyon comes upon in stanza 57, is not an incongruous feature of the cave in light of the statement, again from Biringuccio, that "all those mountains out of which springeth great abundance of water, do also abound with mine."[29] In a passage of considerable interest, Martyr speaks of the advantages of mining on the surface in the New World:

> The Spaniard . . . shall not need hereafter with undermining the earth with intolerable labor to break the bones of our

> mother, and enter many miles into her bowels, and with innumerable dangers cut in sunder whole mountains to make a way to the court of infernal *Pluto*, to bring from thence wicked gold the seed of innumerable mischiefs, without the which notwithstanding we may now scarcely lead a happy life sith iniquity hath so prevailed and made us slaves to that whereof we are lords by nature.[30]

Even here, in what is basically a discourse on the hazards of underground mining, the practical grades into the metaphysical; one digs one's way (through the bowels of one's own mother) not only to Hell but to one's own damnation.

Several aspects of Mammon's Cave lead toward a sense that Spenser's "mine" has a particular locus and is not just a general type. The "deformed creatures, horrible in sight" (7.35.7) who "swincke" and "sweat" (7.36.9) to smelt and cast the metal (ironically, since Mammon has claimed to be above "swinck and sweat" in stanza 8), display an odd reaction to Guyon's appearance:

> But when as earthly wight they present saw,
> Glistring in armes in battailous aray,
> From their whot worke they did themselues withdraw
> To wonder at the sight: for till that day,
> They neuer creature saw, that came that way.
>
> (37.1–5)

These "feends" are innocents; they have never seen a man in armor before, and this accounts for their being dumbstruck. The vividness and foreignness of the scene led one past editor of Book 2 to ask rather naively, "What can Spenser have seen of great furnaces and smelting like this? He seems to anticipate the Industrial Revolution."[31] Spenser is not predicting the future but describing his present-day reality. He has not seen it himself, it is true, but sources of information about it are ready to hand. For instance, Martyr, in one of his more pessimistic moments, describes the forced labor in the New World mines as a kind of living death that reduces its native victims to the level of Spenser's "feends":

> among these so many blessed and fortunate things, this one grieveth me not a little: That these simple poor men never brought up in labor, do daily perish with intolerable travail in the gold mines: And are thereby brought to such desperation, that many of them kill them selves, having no regard to the procreation of children. . . . And albeit that by the kings letters patents it was decreed that they should be set at liberty, yet are they constrained to serve more than seemeth convenient for free men. The number of the poor wretches is wonderfully

> extenuate. They were once reckoned to above twelve hundred thousand heads: But what they are now, I abhor to rehearse. We will therefore let this pass: and return to the pleasures of *Hispaniola*.[32]

Las Casas, unlike Martyr, never lets such matters pass. Appended to *The Spanish Colonie* is a selection from the Twenty Remedies proposed by Las Casas in 1542 during his great debate with Sepulveda concerning the treatment of the Indians; in the Eighth Remedy he contemplates conditions in the mines:

> The labor that they were put unto, was to draw gold, whereto they had need have men of iron. For they must turn the mountains 1000 times upside down, digging and hewing the rocks, and washing, and cleansing the gold it self in the rivers, where they shall continually stand in the water until they burst, and rend their bodies even in pieces. Also when the mines peradventure do flow with water, then must they also besides all other labors, draw it out with their arms. To be brief, the better to comprehend the labor that is employed about gathering gold and silver, it may please your Majesty to consider, that the heathen Emperors (except to death) never condemned the Martyrs to greater torments, than to mining for metal.[33]

In stanzas 35–37, Spenser has literalized the "hellishness" of such contemporary descriptions of the Spanish mines. These grotesqueries are linked in turn with intimations of violence and cruelty on a far greater scale. In the House of Richesse, "all the ground with sculs was scattered, / And dead men's bones, which round about were flong, / Whose liues, it seemed, whilome there were shed, / And their vile carcases now left vnburied" (7.30.6–9). Earlier Guyon has argued that Mammon's plenty leads only to slaughter and disorder:

> But realmes and rulers thou doest both confound,
> And loyall truth to treason doest incline;
> Witnesse the guiltlesse bloud pourd oft on ground,
> The crowned often slaine, the slayer cround,
> The sacred Diademe in peeces rent,
> And purple robe gored with many a wound;
> Castles surprized, great cities sakt and brent:
> So mak'st thou kings, and gaynest wrongfull gouernement.
> (7.13.2–9)

This stanza can be interpreted as a general account of Mammon's depredations throughout history, but Guyon speaks of guiltless blood being shed, of the murder of kings, of "great cities sakt and brent." Spenser seems to indicate here an awareness of the sweeping accusations of Las Casas, of the deaths of Montezuma and Atahualpa, of the destruction of great civilizations at Tenochtitlán and Cuzco. The "wrongfull gouernement" of the New World was stuff of recent vintage, exotic and memorable, and available in a variety of forms to a learned and imaginative English reader.

The references to "guiltlesse bloud pourd oft on ground" and in stanza 30 to "vile carcases" that seem to have been slain on the spot and are now "left vnburied" suggests an underlying association between the forces opposing Guyon and the "pagan" acts of human sacrifice and idolatry. This association is partly embedded in the word "Mammon" itself, which, as the *Oxford English Dictionary* explains it, comes to refer particularly to wealth conceived as an idol to be worshiped. Mammon calls himself a god (7.8.1), and Guyon addresses him more pointedly as a "Money God" (7.39.1). The canto also contains other images of idolatry less closely tied to Mammon and money. Spenser compares the court of Philotime to a "solemne Temple" as well as a "Gyeld" (43.4), and Philotime herself appears like an idol:

> Her face right wondrous faire did seeme to bee,
> That her broad beauties beam great brightnes threw
> Through the dim shade, that all men might it see:
> Yet was not that same her owne natiue hew,
> But wrought by art and counterfetted shew,
> Thereby more louers vnto her to call.
>
> (7.45.1–6)

Philotime represents a parody of the sun-likeness recurrently attached to the various personifications of Elizabeth in Book 2. Bernhart sees an additional dimension to Philotime's court in that Philotime holding the "great gold chaine" of Ambition bears a loose resemblance to the sacred sun images of the Incas.[34] One would not want to press this analogy too far, but it emphasizes again the way in which Spenser exposes the essentially barbarous quality behind apparently "civilized" activity. Climbing the chain of ambition is in itself an act of violence and cruelty, for "euery one did striue his fellow down to throw" (7.47.9). Spenser stresses the kinship of idolatry and violence by placing Disdain inside the gate to Philotime's chamber. Disdain, "full of stomacke vaine" and driven by his Titanic lineage to "scorne all creatures great and small, / And with his pride all others powre deface" (7.41.3, 7–8), is "all of golden mould" (40.7), suggesting both the source of his pride and power and the fact that Disdain's own appearance resembles that of an idol. He is also a blasphemer, "striding stiffe and bold / As if the highest god defie he would" (7.40.4–5). Spenser concludes that Disdain is "More fit amongst blacke fiendes, then men to haue his place" (7.41.9). This judgment would place Disdain among the deformed laborers in stanzas 35–37.

Through his depiction of figures like Philotime and Disdain, Spenser weaves a network of corollary evils surrounding idolatry, a network that traps men and draws them into sacrilege and brutality, transforming them into savages. The final expression of this savagery is human sacrifice ("dead mens bones . . . whose liues . . . whilome there were shed"), which may help to account for the unusual pairing of Tantalus and Pilate in stanzas 57–62. Both are, in a sense, slaughterers of innocents—Tantalus of his own son, Pilate of God's son. Tantalus bears the additional onus of cannibalism, and resembles Disdain in "Accusing highest *Ioue* and gods ingrate / And eke blaspheming heauen bitterly" (7.60.7–8). Yet Tantalus is a man capable of serving a dinner acceptable to the gods and Pilate is the viceroy, so to speak, of a great empire. Both represent the perversion and negation of prosperity and authority among apparently civilized men.

The presence of Tantalus and Pilate in canto 7 thus contributes in a small way to a vision of the Spaniard as more pagan than the primitive nations over which he rules. His worship of Mammon is damning evidence of his paganism, for he has knowingly embraced one of the idols of the natives he is ostensibly trying to convert to true religion. Mammon is treated as such an idol in Nicholas's introduction to his translation of Gómara:

> In this history doth appear the simplicity of those ignorant *Indians* in time past, yea and how they were deluded in worshiping Idols and wicked *Mammon,* their bloody slaughter of men in sacrifice, and now the great mercy of Jesus Christ extended upon them in lightening their darkness, giving them knowledge of the eternity, and holy trinity in unity, whereby they are now more devout unto heavenly things than we wretched Christians, (who presume of ancient Christianity), especially in Charity, humility, and lively works of faith.[35]

Nicholas's use of "Mammon" is linked not only with idols but with human sacrifice. Its definition as "money" or "riches" becomes secondary, even tertiary; New World peoples were not commonly believed in Nicholas's time to possess the Old World vice of greed. Another interesting feature of the passage is its description of an inversion in the social and spiritual order: the Indians have become better Christians than their converters. Las Casas offers an emblematic tale of one such colonial "converter" literally turning to idols rather than to God at the moment of death:

> One of these abominable chafferers, named *John Garcia,* being sick, and near his death, had under his bed two packs of Idols, and commanded his Indish maid that served him, to look to it that she made not away his idols . . . for they were good stuff: and that making vent of them, she should not take less than a

> slave for a piece one of them with another: and in fine, with this his Testament and last will thus devised, the caitiff died, busied with this deep goodly care, and who doubteth but that he is lodged in the bottom of hell[?][36]

Here the different sixteenth-century meanings of Mammon come together, for Garcia worshiped his idols insofar as they were worth something to him. What they were worth to him was slaves, and in the exchange of idol and slave, "one of them with another," can be seen again the potent analogy between slavery and human sacrifice. Las Casas's sarcasm is proportionate to his sense of the enormity of Garcia's blasphemy—a blasphemy despite the fact that Garcia's "last will" was also an act of generosity toward his Indian maid or, more likely, mistress.

Gold-hunger, slavery, treachery, cruelty, sacrilege—these are the landmarks of the literary terrain devoted to the Conquista, and it is into this terrain that Guyon descends in canto 7. The descent has usually been construed as a series of temptations that Guyon more or less successfully resists by virtue of his temperance. Yet "temptation" may be a misleading term for what Guyon undergoes, since it implies a more intense level of activity than transpires in the canto. "Temper" has always been mainly a transitive verb, something that one does to something else—to glass, to a sword, to one's own feelings and desires. Under this qualification, Guyon does little "tempering" in canto 7. In fact, he does little of anything, in contrast to the immense and even frenzied activity going on around him during his sojourn in the cave. His response to Mammon's various snares is placid, to say the least. He shows a slight, though perhaps significant, deference to Philotime ("Gramercy *Mammon* . . . / For so great grace and offred high estate" [7.50.1–2]); this passes quickly. He comes close to doing battle with Disdain in stanza 42, but Mammon forestalls him. The attraction of the golden apples and silver stool in the Garden of Proserpina appears to be lost on him—not surprisingly, considering Mammon's rather comical lack of subtlety in tempting Guyon with them: "Thou fearefull fool, / Why takest not of that same fruit of gold, / Ne sittest downe on that same siluer stoole" (7.63.6–8). When Tantalus pleads with him to "giue to eate and drinke to mee" (7.59.9), Guyon does nothing but advise him on the error of his ways.

Most of Guyon's acts are speech acts, thus placing them in a marginal category. They do not infringe on the world around him, and thus they verge on the passive. This passive quality affects his behavior throughout canto 7; his main activity, besides speaking, is watching. Though he moves from one end of the cave to the other, this is an act of following dependent upon Mammon's presence. Even when Guyon moves, his passive character is maintained.

All of this leads to the proposition that Guyon is best understood not as a subject to be tempted but as an object to be tested. "Test" is the operative word here: Guyon is presented less as a being with his inner life on trial than as a kind of opaque substance undergoing an examination of its essential qualities. The test, reduced to its crudest formulation, is to determine whether he is a Spaniard or not. Spenser has carefully devised a setting that refers in numerous oblique ways to the Spanish New World; he places Guyon in this setting and lets its pressures work upon him. The outcome of the test is negative from the beginning—Guyon enters the cave without his horse or the Palmer—but negative in a circumscribed way. There are superficial similarities between Guyon and the Spaniards; like them, he suffers from a real hunger resulting from his ordeal in the "mines." Both Berger and Cain have claimed that Guyon also suffers from, in Berger's words, "an unprofitable curiosity," which partly accounts for his faint at canto's conclusion.[37] This is the generic curiosity of the explorer trying to comprehend the foreign and unknown, a curiosity common to Englishman and Spaniard alike, and not ordinarily condemned in the sixteenth century. The Englishman must enter the New World much as the Spaniard has entered it before him and experience many of the same rewards and hardships; but he does so, Spenser would have his readers believe, for different reasons.

I return to Eden's statement of the simple truth that gold is a means and not an end, to be used "as do cunning physicians" to achieve a higher purpose. For Spenser, as for Eden, that higher purpose is the acquisition of power. Guyon displays his knowledge that gold is a means rather than an end when he asks Mammon why he hides his wealth "apart / From the worldes eye, and from her right vsaunce?" (7.7.3–4). He also implies by this question that he has a correct understanding of what the "right vsaunce" of wealth in the world would be—namely, the enhancement of power. Guyon demurs from accepting Mammon's offer in stanza 9 by saying, "Regard of worldly mucke doth fowly blend, / And low abase the high heroicke spright, / That ioyes for crownes and kingdomes to contend" (7.10.5–7). This is a striking assertion on Guyon's part. His intentions go further than "Faire shields, gay steedes, bright armes" (7.10.8), a fact that comes through still more clearly in Guyon's response to viewing the House of Richesse:

> To them, that list, these base regardes I lend:
> But I in armes, and atchieuements braue,
> Do rather choose my flitting houres to spend,
> And to be Lord of those, that riches haue,
> Then them to haue myself, and be their seruile sclaue.
> (7.33.5–9)

Guyon may be speaking of lordship metaphorically, as an aspect of moral or spiritual life; yet it is hard to avoid seeing this speech as possessing a blunt political meaning. Guyon wishes literally "to be Lord of those, that riches haue." His sense of his larger mission, in which gold is a mere collateral attraction, enables him to say to Mammon a few stanzas later, "giue me leaue to follow mine emprise" (7.39.6). Guyon's affinity for power also accounts for his politeness toward Philotime, whose golden chain covered with struggling votaries most closely resembles the struggle for power seen as distinct from the hunger for wealth.

What does it mean, though, to say that Guyon hungers after power? More importantly, what does it mean in the historical context? Greenblatt claims of Guyon's creator that "Even when he most bitterly criticizes its abuses or records its brutalities, Spenser loves power and attempts to link his own art ever more closely with its symbolic and literal embodiment."[38] Greenblatt finds that embodiment in the court and cult of Elizabeth, and on a more general level in the emergent imperialism he posits within Elizabethan culture; it is the hungering after this sort of hegemonic power that culminates in the destruction (read repression) of the Bower of Bliss in 2.2.83. As influential as Greenblatt's interpretation has proved in the last twenty years, it is in a way too "timeless" as a description of the Elizabethan desire for power, at least as far as that desire is demonstrated in canto 7. The impetus "to be Lord of those, that riches have" had a specific meaning for the political and commercial entrepreneurs who promoted English efforts at expansion overseas; Drake, Ralegh, and their supporters believed that by co-opting the wealth of the New World—whether by piracy or by settlement, or even by collecting customs duties from Spanish shipping at strategically located fortified ports—England would eventually supplant Spain as the first nation of Europe.

Among many other schemes in the *Discourse of Western Planting*, Hakluyt proposes that the English foment rebellion among the Indians as a way of weakening Spanish control of the Indies:

> And like as the foundation of the strongest hold undermined and removed, the mightiest and strongest fall flat to the earth, so this prince spoiled and intercepted for a while of his treasure, occasion by lack of the same is given that all his Territories in Europe out of Spain slide from him, and the Moors enter into Spain it self, and the people revolt in every foreign territory of his, and cut the throats of the proud hateful Spaniards their governors.[39]

To wish for the return of the Infidel to Europe seems extreme, however humiliating it might be for Philip, but Hakluyt hopes for an equally important change in England's status as well. When he suggests aiding rebellion among the *Cimarrones* or Maroons—African slaves who had intermarried with Caribbean tribespeople—he states the goal rather modestly: "by this means, or by a platform well to be set down, England may enjoy the benefit of the Indian mines, or at the least keep Philip from possessing the same."[40] Later in the manuscript, Hakluyt is more forward: "Their manifold practices to supplant us of England give us most occasion to bethink ourselves how we may abate and pull down their high minds."[41] Hakluyt plays this theme in multiple variations throughout the *Discourse;* acquiring power in the New World is ultimately acquiring power over Spain, and the resources of the New World—as employed by England's cunning physicians—are the means to that end. If Guyon is viewed as Spenser's portrait of the anti-conquistador, confronting the vices of the Spanish New World from a peculiarly English perspective, then one can conjecture that Guyon's "emprise" is the pursuit less of *an* empire than of the *Spanish* empire, or rather of all of the implications that this empire carried in the Elizabethan cultural imagination.

Presumably for Spenser, Guyon's "lordship" would plant "right vsaunce"—whether in government, in religion, or in commerce—in place of all that is wrong in Mammon's kingdom. Yet such lordship verges on paradox, since it would also supplant what is presently "higher up," so to speak, and thus bears an uncomfortable resemblance to the activity on Philotime's chain, where "Those that were vp themselues, kept others low, / Those that were low themselues, held others hard, / Ne suffred them to rise or greater grow" (7.47.6–8). Guyon declines Mammon's offer of Philotime's hand in marriage, pleading both incompatibility (he is mortal, she is immortal) and a prior allegiance: "yet is my trouth yplight, / And loue auowd to other Lady late, / That to remoue the same I haue no might" (7.50.6–8). Yet Guyon's "trouth" to this unnamed other lady, whatever she represents, is not entirely free from the savor of the ambition—and idolatrousness—that condemns Philotime's minions to their struggle on the chain.[42]

The conclusion that Guyon, in the name of "right vsaunce," might be climbing his own golden chain (at the top of which, curiously enough, Philip II sits enthroned), is surely not the one that Spenser wishes the reader to reach; he defers it by allowing Guyon's character to hover ambiguously between the poles of the canto's argument. Guyon's desire for lordship may place him ahead of the typical conquistador, who is terminally stalled at the desire for wealth; yet at the same time, Spenser wants to position Guyon behind the Spaniard, in a state of primitive, untainted goodness—in other words, in a state like the native's. Guyon's praise of "Vntroubled Nature" and the "antique world" (7.15.4, 16.1) suggests that his moral status is not far from that of Comogrus's son when the latter admonishes the Spaniards in Martyr's text. Guyon places himself among the natives in speaking of the "superfluities" that "empeach *our* natiue ioyes" (7.15.5, 6; my emphasis). Thus Guyon participates in the inversion of native and conquistador that occurs frequently in the New World literature and places the Indian in a position of moral, even spiritual, superiority to the Spaniard.[43]

This identification of Guyon with the Indian can only go so far. While Comogrus's son is admonishing the Spaniards, he regrets that his own countrymen "destroy one an other by reason of ambition and desire to rule." The Indians, it seems, are no strangers to the vices of their oppressors; if Indians cannot claim immunity to the disease of ambition and its attendant complications, Englishmen can much less. It is therefore crucial that Guyon's character never receives a positive and conclusive demonstration in action in canto 7. For the values that serve to distinguish Guyon from the Spanish model have a way of collapsing, if pressed too hard or allowed to sink under the weight of their own internal contradictions, into similitudes with that same model; this is perhaps the greatest peril in Guyon's whole "emprise." Thus, Guyon's faint at the end of the canto is a kind of rescue in itself, above and beyond the angelic assistance he receives at the beginning of canto 8. When Guyon loses his senses, "as ouercome with too exceeding might" (66.7), he escapes from the looming consequences of having become what I termed him earlier—a Spaniard immaculately conceived. Guyon's faint also serves to dissipate, or at least postpone, any sense that Guyon personifies Spenser's concrete vision of an English empire in the New World. Insofar as that vision exists, it is still predicated upon distancing the Englishman from the Spaniard. But the Spaniard remains the standard by which the distance is judged; the British empire has, in a sense, not yet passed beyond the Pillars of Hercules into the Ocean Sea. Even so, to say that Spenser purposely fails to define Guyon's identity would be to overstate the case; what Spenser does, rather, is initiate a process of definition that never arrives at a conclusion. It is worth recalling in this context that the verb "temper" can refer not only to the hardening of an object but also to its softening or melting. Guyon, like so much of the gold in Mammon's Cave, and like so many of the other figures in Spenser's protean poem, is being tempered in just such a way; he is molten, malleable, carrying the potential but not necessarily the actuality of a stable and eternal form.

CHAPTER 5

Gnats, Mantles, Tigers, Shades: Maleger and the "Irish Question"

In the long traverse between the Cave of Mammon and the Bower of Bliss, the House of Alma is the most significant landmark. It is significant, of course, as an allegorical representation of the temperate human body, and as the repository of the Briton and Elfin chronicles. But one of the most memorable features of Alma's castle is that it is under siege. My concern in this chapter is with the assault on the castle by the cryptic figure of Maleger and his "raskall routs" (9.15.4), which is broached in canto 9 and continues during Arthur's battle with Maleger in canto 11. I will argue for the basic consistency of the historical dimension in Book 2, and the limited utility of "reading Ireland" into its last cantos if not into the book as a whole.

Some bold historical analogies have been advanced over the years based on the notion that Book 2 offers allegorical reflections of Spenser's Irish experience: the hazards of Guyon's voyage in canto 12 have been equated with the coastal geography of County Kerry; Maleger has been compared to the Ulster rebel Shane O'Neill; Pyrocles and Cymocles have become variants on characters in the Gaelic tale of the battle between Cuchulain and Ferdiad, with Atin substituting for Cuchulain's dart-carrying servant Laeg.[1] Such analogies are grounded on the assumption that Ireland's presence in Book 2 is a given. Critical faith in this assumption has shown up most obviously in the effort to interpret the Maleger passages. The description in canto 9 of the "Vile caytiue wretches" outside of Alma's castle (9.13.4)—the group identified two cantos later as Maleger's army—prompts one editor of Book 2 to say that "clearly Spenser is drawing here on actual experience in Ireland, and not merely on romantic and narrative precedent."[2] This generalization appears to follow in the wake of remarks by earlier twentieth-century readers of Spenser such as Pauline

Henley and M. M. Gray, the latter of whom saw an analogy between the "thousand villeins" of cantos 9 and 11 and the Irish who participated in the burning of Spenser's estate at Kilcolman.[3] Even Greenblatt, whose main interests lie elsewhere, assumes the "Irishness" of Book 2 in his reading of Spenser, claiming that "Ireland is not only in Book 5 of *The Faerie Queene;* it pervades the poem." This claim leads to a remark not unlike those once made by Gray concerning Book 2: "[Spenser's] imagination is haunted by . . . nightmares of savage attack—the 'outrageous dreadfull yelling cry' of Maleger."[4] The first of the two quotations from Greenblatt hints at the problem that I want to address here at greater length than in the introduction. There are points in *The Faerie Queene,* notably in Book 5, where the matter of Anglo-Irish relations in the sixteenth century seems inseparable from any discussion of the poem's meaning. My rather literal-minded argument is that for Book 2, the matter of Ireland does not provide a reliable medium of interpretation; the Maleger material remains congruent with the reading of the poem that has been outlined in the previous chapters.

Confidence in the notion that canto 11 allegorizes the Elizabethan Anglo-Irish conflict is largely prompted by the fact that Spenser introduces Maleger's mob with, in Gray's words, "a simile always pointed out as the first allusion to Ireland in the *Faerie Queene.*"[5] This is the extended simile in canto 9 stanza 16 that begins, "As when a swarme of Gnats at euentide / Out of the fennes of Allan do arise" (9.16.1–2). The difficulty with using this passage as the basis for an interpretation is, as usual, the polysemous character of Spenser's analogies (I am well aware that in preparing the argument of this book I have had to deal, I hope responsibly and productively, with the same difficulty). In canto 11, for instance, Spenser compares Maleger's painted arrows to "Such as the *Indians* in their quiuers hide"; like the Indians' arrows, these are poisonous: "Ne was their salue, ne was their medicine, / That mote recure their wounds: so inly they did tine" (11.21.5, 8–9).[6] Later he describes Maleger's style of warfare—firing arrows in retreat—in very different terms: "in his flight the villein turn'd his face, / (As wonts the *Tartar* by the *Caspian* lake, / When as the *Russian* him in fight does chace)" (11.26.6–8). From these allusions, Maleger and his forces could be identified as readily with Indians or Tartars as with the Irish. All three identifications can be subsumed into a general category of barbarism, but the impact of "the first allusion to Ireland in the *Faerie Queene*" is such that the "Irishness" of Maleger has received the most emphasis. What the Ireland-oriented readers of Book 2 also overlook is that while "the fennes of Allan" may represent the first Irish reference in the poem, it is one of only three such references in Book 2, none of which are closely related. The second occurs in canto 9 stanza 24 and describes the stone in the porch of Alma's castle: "Stone more of valew, and more smooth and fine, / Then Iet or Marble far from Ireland brought" (9.24.2–3); it is worth noting that the stone has been brought *from* Ireland, indicating that Alma's castle is elsewhere. The third reference is part of the "chronicle of Briton kings" in canto 10: "[Gurgunt] also gaue to fugitiues of *Spayne* . . . A seate in *Ireland* safely to remayne" (10.41.6, 8). References to Spain in Book 2 indeed raise other issues, some of which I have already tried to illustrate.

The remainder of this chapter briefly examines some of the problems created by an "Irish" interpretation of Maleger, and point to ways in which these same questions can be handled at least equally well under an "American" interpretation. I will conclude with some remarks on the relationship between Arthur's role in canto 11 and Guyon's in canto 12; since Arthur is the perfect knight, embodying temperance as well as all the other virtues, his actions offer both an iteration of and a counterpoint to Guyon's own at the end of Book 2.

The mysterious figure of Maleger presents an instructive instance of how readily *The Faerie Queene*'s topicality can be rendered in terms of the immediate contexts of Spenser's career, but also of how easily the personal-historical argument (by which I mean an argument based on a more flexible frame than that offered by traditional biography) can become reductive or misleading. A case in point would be the provenance of Maleger's "armor," described in canto 11 as follows:

> All in a canuas thin he was bedight,
> And girded with a belt of twisted brake,
> Vpon his head he wore an Helmet light,
> Made of a dead mans skull, that seem'd a ghastly sight.
> (11.22.6–9)

Gray found confirmation of Maleger's identity in the remarks on the Irish mantle in *A View*,[7] where Spenser describes this native garment as "never heavy, never cumbersome" and useful to rebels because "it is light to bear, light to throw away, and being as they then commonly are naked, it is to them all in all."[8] In other respects, however, the mantle in *A View* seems less than a close match for Maleger's garment. The prevailing impression of this garment is that it is flimsy, appearing like "graue-clothes vnbound" and thus well-fitted to Maleger's "leane and meagre" body (11.20.9, 22.2), whereas Spenser emphasizes the mantle's sturdiness. Arguing for its origin among the Scythians, who had allegedly invaded Ireland in the distant past, Spenser's spokesman Irenius claims that "it was renewed and brought in again by those northern nations, when breaking out of their cold caves and frozen habitation into the sweet soil of Europe . . . and coming lastly into Ireland they found there more special use thereof, by reason of the raw cold climate, from whom it is now grown into that general use in which that people now have it."[9] The mantle is tough enough to be used as a shield:

"being wrapped about their left arm instead of a target . . . it is hard to cut through it with a sword."[10] Neither of these observations fits with the description of Maleger's "canuas thin."

Maleger's garment does, however, bear a resemblance to the woven battle gear worn by various tribes in the Americas[11]—suggesting that *A View* is only one of several kinds of texts that can be brought to bear on Maleger's description. Martyr, for instance, narrates the encounter of Rodrigo de Colmenares's party with "a certain King . . . appareled with vestures of gossampine cotton, having twenty noble men in his company appareled also . . . the Kings apparel, hung loose from his shoulders to his elbows: And from the girdle downward, it was much like a womans kirtle, reaching even to his heels." The king and his men prove to be more dangerous than they at first appear, "so fiercely assailing our men with their venomous arrows, that they slew of them forty and seven before they could cover them selves with their targets. For that poison is of such force, that albeit the wounds were not great, yet they died thereof immediately. For they yet knew no remedy against this kind of poison."[12] There is also no remedy for the arrows in Maleger's quiver.[13] What these arrows do not resemble are the "short bearded" ones mentioned in *A View*'s description of Irish weaponry, arrows "tipped with steel heads made like common broad arrow-heads, but much more sharp and slender."[14] Maleger's arrows are closer to the pre-Columbian variety, "Headed with flint, and feathers bloudie dide" (11.21.4).[15]

More than any of his other attributes, though, the "Tygre swift and fierce" (11.20.4) which Maleger rides into battle forces readers to move beyond the interpretive matrix provided by Spenser's Irish experience. Tigers, like snakes, do not figure conspicuously in Ireland's natural history. On the other hand, Spenser could be using the tiger for its symbolic associations, or as a generalized marker of a barbarism more eastern than western: the "other" Indians, of the subcontinent, allegedly used tigers as mounts, and such an analogy would complement Spenser's allusion in stanza 26 to the Tartars, who belonged to another part of the East.

However, it is worth noting that early European commentators on the New World showed considerable interest in and anxiety about large American cats (jaguars, pumas, etc.), which were classified under the limited taxonomy of the Renaissance as tigers.[16] Martyr speaks of such a tiger slain in a staked pit: "Being yet dead, he was fearful to all such as beheld him: what then think you he would have done being alive and loose."[17] Among the excerpts from Oviédo's *Historia general y natural de las Indias* that accompany Eden's translation of the *Decades*, there is a passage in which the scientifically minded Oviédo expresses his doubts as to whether these New World animals have been named correctly: "In the firm land are found many terrible beasts which some think to be Tigers. Which thing nevertheless I dare not affirm, considering what authors do write of the lightness and agility of the Tiger, whereas this beast being other wise in shape very like unto a Tiger, is notwithstanding very slow."[18] But he is in no doubt about the fearsomeness of this animal: "I have seen some of three spans in height, and more than five in length. They are beasts of great force, with strong legs, and well armed with nails and fangs which we call dog teeth. They are so fierce that in my judgment no real lion of the biggest sort is so strong or fierce. Of these, there are many found in the firm land which devour many of the Indians and do much hurt otherwise."[19] Though Oviédo's tiger is considerably slower than Spenser's, it is fully as terrible as Maleger himself, and would make an appropriate mount for this frighteningly primitive warrior. This reading is admittedly tenuous—I have found no actual allusions to American natives riding tigers—but it at least makes better contextual sense of Maleger's tiger than an Irish reading does.

A difficulty that arises with any contextualization of material like this from *The Faerie Queene* concerns the poem's typically slippery metaphorical relationships. The animal imagery here is reversed in canto 9, when Arthur and Guyon first do battle with Maleger's rabble: "Those Champions broke on them, that forst them fly, / Like scattered Sheepe, whenas the Sheapheards swaine / A Lyon and a Tigre doth espye, / With greedy pace forth rushing from the forest nye" (9.14.6–9). In this stanza Spenser's two heroes become the predators. Arthur's escape from his temporary captivity in canto 11 is described in even more "bestial" terms:

> . . . [he] broke his caitiue bands,
> And as a Beare whom angry curres haue touzd,
> Hauing off-shakt them, and escapt their hands,
> Becomes more fell, and all that him withstands
> Treads down and ouerthrowes. . . .
> (11.33.2–6)

Yet when Huddibras and Sans-Loy attack Redcross in canto 2, Spenser designates the villains with the same terms:

> As when a Beare and Tygre being met
> In cruell fight on lybicke Ocean wide,
> Espye a traueiler with feet surbet,
> Whom they in equall pray hope to deuide,
> They stint their strife, and him assaile on euery side.
> (2.22.5–9)

These apparent contradictions in imagery can be dealt with in two ways. The first is to attribute them to the sheer conventionality of the images: all of the similes above are rooted in commonplaces, and so may be treated as basically neutral, used by Spenser merely to reinforce a particular effect. The second, and possibly more instructive way, is to examine the hierarchical quality of the similes. In the Renaissance conception of the natural order, lions and tigers are superior to sheep; bears are superior to dogs; and human travelers are superior to bears and tigers. Each simile exposes the moral infrastructure of *The Faerie Queene.*

The same sort of hierarchical metaphors crop up in sixteenth-century voyage literature, usually in the context of religious conversion. In the panegyric to Philip that prefaces the *Decades,* Eden praises the Spanish king's mercy toward the Indians and at the same time distinguishes between lions, lambs, horses, and tigers (here equated with dragons): "Being a Lion he behaved him self as a lamb, and struck not his enemy having the sword in his hand. Stoop England stoop, and learn to know thy lord and master, as horses and other brute beasts are taught to do. Be not indocible like Tigers and dragons, and such other monsters noyous to man kind."[20] The distinction is not only between docible and indocible beasts, but between the humane, civil, Christian lion and the inhumane, uncivil, pagan tiger. If the distinction in imagery is transferred to canto 11, then the battle between Arthur (the future king) and Maleger becomes a fight between a lion and a tiger in which the tiger is ultimately domesticated. A passage from the *Decades* may also be relevant here: the King of Margarita is converted "from a cruel tiger to one of the meek sheep of christs flock sanctified with the water of baptism with all his family and kingdom."[21] This image of conversion fits rather strikingly with A. S. P. Woodhouse's theory in his famous essay on *The Faerie Queene* that Arthur's final defeat of Maleger, achieved by throwing him into "a standing lake" (11.46.6), "is intended to suggest baptismal regeneration, that is to say, it moves in the same area of symbolism as does the sacrament of baptism."[22]

The metaphor also appears, in a formulation fairly close to that of canto 9 stanza 14, in Las Casas's *Brevissima Relación,* where the emphasis is on the repression of the Indians rather than their conversion: "Upon these lambs so meek . . . entered the Spanish . . . as wolves, as lions, & as tigers most cruel of long time famished."[23] This at first seems an ironic anticipation of the metaphor's use in canto 9, except that there is the possibility that Arthur and Guyon act as "undoers" of a whole notion of empire; they renovate and purify not only the actions of imperial Spain, but also the metaphors applied to those actions. Arthur and Guyon, instead of slaughtering helpless innocents, are defending the structure of well-being and civility—Alma's castle—against the forces of corruption and chaos.

A third problematic feature of canto 11, one that is much more abstract in character, involves Maleger's seeming invulnerability:

> Flesh without bloud, a person without spright,
> Wounds without hurt, a bodie without might,
> That could doe harme, yet could not harmed bee,
> That could not die, yet seem'd a mortall wight,
> That was most strong in most infirmitee.
>
> (11.40.4–8)

Spenser is constructing a mystifying paradox here; the paradox is difficult to interpret in any case, but perhaps especially in light of efforts to read the passage within a specific historical-allegorical context. To integrate Maleger's peculiar character with the Irish reading, one would likely have to assume that Spenser is referring to the persistence and resilience, rather than the immortality, of the Irish rebels. This only seems to unravel a few strands of the knot, however. I am unable to propose a definitive explanation for this puzzling passage in canto 11, but would like to offer an analysis that fits well with an American reading and may open up broader avenues of historical interpretation.

Maleger shares his physical qualities with his followers, "idle shades" who "though they bodies seeme, yet substance from them fades" (9.15. 8, 9; once again Spenser privileges the word *idle*). The siege of Alma's castle opens with a salient fact about Maleger's forces: "So huge and infinite their numbers were, / That all the land they vnder them did hide" (11.5.6–7). As "infinite" as these forces are, Spenser indicates that "euery one did bow and arrowes beare" (11.8.7). That the weapons are as numberless as the attackers helps to make more concrete the Virgilian metaphor that Spenser employs to describe the "fluttring arrowes" that fall on Arthur "thicke as flakes of snow, / And round about him flocke impetuously, / Like a great water flood" (11.18.2–4).

It is difficult to reconcile this imagery with the actual circumstances of Irish resistance in the sixteenth century. The rebels' successes owed more to the country's rugged terrain than to numerical strength; moreover, Ireland remained quite sparsely populated relative to England and the rest of northern Europe until well into the eighteenth century. Spenser's description of Maleger's army fits well, however, with contemporary accounts of the population of the New World. Sixteenth-century discussions of native peoples repeatedly stress their sheer numbers—a way of underlining both the awesome promise and the mind-boggling difficulty of the colonial enterprise. In the *Brevissima Relación,* Las Casas speaks of the Caribbean islands as "the most peopled, and the fullest of their own native people, as any country in the world may be." Yet the islands are only microcosms of the American mainland: "It seemeth that God hath bestowed in that same country, the gulf or the greatest portion of mankind."[24] Hakluyt in the *Discourse of Western Planting* resorts innocently to such superlatives as he claims the necessity of "reducing of infinite multitudes of these simple people that are in error into the right and perfect way of their salvation."[25] Las Casas and Hakluyt each represent in his own fashion the promise of the New World and its (allegedly) huge population.[26] The benefits of such a population—providing as it would countless laborers, neophyte Christians, and Spanish or English proxies—can just as readily be inverted into the threat of massive native armies in perpetual descent upon small, doomed bands of voyagers. Gómara's *Istoria de las Indias* vividly depicts such a threat in its account of the events leading up to the infamous *Noche Triste* of July 10, 1520, when Cortés was forced to retreat from Tenochtitlán. Shortly before Montezuma's mysterious and possibly accidental death, the Aztecs "beseiged the Spaniards house, with such strange noise that one could not hear another: the stones flew like hail, Darts and arrows filled the Spaniards yard, which troubled them much."[27] The resistance mounted by the Spaniards led to disheartening results: "if a shot carried ten, fifteen or twenty *Indians* at a clap, they would close again as though one man had not been missing. . . . the *Indians* were so many in number, that no hurt appeared, yea and our men were so few in comparison of them, that although they fought all the day, yet had they much a do to defend themselves, how much more to offend."[28] During Cortés's retreat in darkness across the city's bridges, the Aztec attack began to take on apocalyptic proportions: "they having no armor to put on, nor other impediment, joined an infinite company of them together, and followed with great celerity, yea and with such a heavy and terrible noise, that all the lake pronounced the Echo, saying, let the cursed and wicked be slain, who hath done unto us such great hurt."[29] This near-conclusive disaster did not deter Cortés for long, but the imagery attached to it soon entered the consciousness of those who, as actors, chroniclers, or theorists, followed him in the project of conquest.

These images also suggest a possible intertextual explanation of Maleger's ability to "himselfe to second battell bend, / As hurt he had not bene" (11.35.5–6), of why "Ne drop of bloud appeared shed to bee / All were the wounde so wide and wonderous," and of why when Arthur "made his spright to grone full piteous: / Yet nathemore forth fled his groning spright, / But freshly as at first, prepard himselfe to fight" (11.38.1–2, 7–9). Maleger, along with his followers, personifies not only the threat of disease to the body natural and savagery to the body politic, but also the threat of confrontation with an innumerable, impersonal enemy on a battlefield without known boundaries.[30] This enemy uses to advantage the ground which has nurtured him: "th'Earth his mother was, and first him bore; / She eke so often, as his life decayd, / Did life with vsury to him restore, / And raysd him vp much stronger then before" (11.45.2–5). Seeing Maleger as a figure of human peril, albeit such peril conceived on a grand scale, helps to situate his resemblance to Antaeus within a wider perspective than could be gained simply from studying Spenser's use of classical mythology or, for that matter, Spenser's tenure in Ireland. Seeing him in this way may also lead to a heightened appreciation of the stratagem that Arthur uses to defeat Maleger at last: he removes him from the land and casts him into the water, thus performing a symbolic act of dispossession. In the history of conquest and colonization, and for either side in a particular colonial conflict, dispossession is nearly always equivalent to defeat. Interestingly, too, Arthur achieves his earlier rout of Maleger's army in canto 11 with one of the chief weapons used by the conquistadors to defeat numerically superior forces—the horse: "And vnder neath him his courageous steed, / The fierce *Spumador* trode them downe like docks" (11.19.6–7).

Guyon and the Palmer depart on their voyage to the Bower of Bliss in the fourth stanza of canto 11, just as Arthur's battle with Maleger is about to begin. Spenser thus indicates that Arthur's and Guyon's ordeals are synchronous—perhaps, in a sense, part of the same struggle. Though Guyon must travel farther than Arthur in his quest, Arthur is in no sense "at home"; both protagonists are fighting on foreign shores, but on different fronts. This is a place where the Irish reading might plausibly settle itself: Guyon sails to the New World, while Arthur remains behind to put down the kerns. Yet the details of canto 11 as discussed above would seem to preclude such an account. It would seem, too, that there might be some sort of retrospective development of the Irish theme in canto 12 if this were indeed crucial to Spenser's design. Ireland is not on Spenser's mind in any obvious way in the concluding canto, however, and the efforts of critics to locate Ireland there have been sketchy at best.[31]

At the same time, Spenser provides a symmetrical arrangement of themes in the two cantos, which are well-suited to an interpretation that reaches toward New World horizons. Maleger's daunting ephemerality is not unlike Acrasia's elusiveness; Guyon must bind her "in chaines of adamant . . . / For nothing else might keepe her safe and sound" (12.82.6–7). In both instances the protagonist encounters a foe who is, literally and metaphorically, difficult to grasp. More important still is the way in which Guyon's and Arthur's positions dovetail in the last two cantos. I suggested in chapter 3 that Arthur's character is quite different from Guyon's; yet at key points in the poem Arthur and Guyon resemble one another in their dependence. The Palmer's rescue and the revival of Guyon at the beginning of canto 8 leads to Guyon's near total reliance on his companion's directives by the end of the book. Meanwhile, in canto 11 Arthur is being ambushed by Impotence and Impatience, "And of the battell balefull end had made, / Had not his gentle Squire beheld his paine, / And commen to his reskew, ere his bitter bane" (11.29.7–9). Spenser goes on to remark, "So greatest and most glorious thing on ground / May often need the helpe of weaker hand" (11.30.1–2). Stanza 30 serves to advertise not only the power of providential grace but also the considerably more pragmatic power of human cooperation. One aspect of Spenser's notions about colonialism that emerges with relative clarity in Book 2 is that heroes, even Arthurian heroes, do not colonize alone; however prodigious their labors in the vast wild plains, woods, and vales of *The Faerie Queene,* they occasionally need some efficient assistance.

In my discussion here I have not attempted to do full justice to the allegorical complexities of canto 11, especially in its dimension as an account of the external threats to the temperate body as a "fortress of health." A historical reading is necessarily a limited one, providing only a partial understanding of a given text. But some historical readings are more limited, and limiting, than others, and I have aimed to show how the Irish reading is an insufficient aid to understanding this particular canto. Insofar as Ireland has an effect on Spenser's methods in Book 2, it may function as a kind of optical filter through which he views the prospects for an English presence in more distant places. Many scholars have detailed the way in which the Elizabethans' experiences in Ireland conditioned their views of America and native Americans.[32] But this sort of condition is more evident in *A View* than it is in Book 2 of *The Faerie Queene*.[33] The problem in comparing the two works is one to which historically oriented critics are especially prone (and one that is probably not entirely avoided in this book): a desire to substitute what is obvious in factual record—such as Spenser's long residence in Ireland—for what is enigmatic in literature. Yet much of Spenser's oeuvre falls into the latter category, and must be treated with as much respect for its frustrations as for its ultimate rewards. The Maleger episode shows what many other instances in the poem show just as effectively, that the riches of *The Faerie Queene* can never be reduced simply to events, no matter how enormous or trivial, in Spenser's laconic personal history.

CHAPTER 6

The End of All Our Travel: Guyon at the Bower of Bliss

Viewing the Cave of Mammon and the House of Alma as "ports" on Guyon's voyage requires a capacity for inference in the reader that proves unnecessary in canto 12, where the activity of voyaging is a salient feature of the narrative. So accessible is the voyage motif here that Spenser's professional readers tend to leap from the global vantage of the proem directly into the waters surrounding the Bower of Bliss. Indeed, canto 12 offers an embarrassment of riches to readers who interpret Book 2 as a poetic extension of the archetypal Renaissance travel narrative, divided as it is fairly evenly between a perilous marine journey and an exotic island destination at journey's end.

One emphasis in this chapter will be the provisional epistemological character of the journey—a character that can be extended to Guyon's quest as a whole. Spenser is not writing from the assurance of having fixed the boundaries of the *oikumene,* whether one understands that "known world" literally or figuratively. Instead, like a number of his more obviously adventurous countrymen, he is involved in an act of exploration. The world he explores is still a mysterious one in which absolute knowledge (the knowledge on which predictions can be founded) has yet to be gained. In this sense, Spenser is not concerned with setting standards for national activity but in seeking out possible sites on which such standards might be erected. *The Faerie Queene* embodies Spenser's effort to acquire knowledge, not to display what is already known.

This exploratory quality is evident in Spenser's approach to reference in canto 12—an approach both eclectic and dynamic. The narrative content of the voyage to the Bower relies on a wealth of writing on the topic of perilous journeys by sea, both inside and outside of history; it has clear antecedents in the *Odyssey* and the *Aeneid*, in Ariosto and Tasso, and in the traditions of classical Mediterranean geography represented by Herodotus, Pliny, and Strabo. One might be hard put to find a valid place for the New World in this mélange with so many identifiable elements. Yet for Spenser, as for other writers of his generation, the lore of the past is never static but is constantly being appropriated to new, occasionally quite practical, uses. In the geographical literature of the sixteenth century, the terms of ancient myth and history are updated to meet and comprehend the novelties that appear on the horizon with each new voyage. This updating seems to arise not out of some systematic desire to classicize the barely known worlds beyond the sea, but out of a desire to make these worlds safer and more accessible. If a danger is similar to one already encountered closer to home, then it is easier to confront in its new form. As Elliott observes, "It is hard to escape the impression that sixteenth-century Europeans . . . all too often saw what they expected to see [in the New World]. . . . [I]t may well be that the human mind has an inherent need to fall back on the familiar object and the standard image, in order to come to terms with the shock of the unfamiliar."[1] Elliott is addressing the negative consequences of this reliance on old models, especially in regard to the treatment of native populations. On the other hand, such a reliance has a fairly harmless side, which manifests itself in the conventions of voyage literature: the writers of such literature often provide classical "handles" that allow readers to grasp the wonders gradually being opened before them.

The Rock of Vile Reproach and the Gulf of Greediness in the early stanzas of canto 12 bear a clear relation to Scylla and Charybdis in the *Odyssey*. This relation becomes more complex, however, in light of the fact that voyage chronicles of the mid-sixteenth century also take advantage of the similarity between Homer's perils and those facing the Spaniards in the New World. Martyr writes of a dangerous place in the Gulf of Uraba encountered by Martín Fernández de Enciso, mainly noted as the author of *Suma de Geographia* (1519) but also the leader in 1510 of an effort to settle Darien: "The flowing of the sea raged and roared there, with a horrible whirling as we read of the dangerous place of Scylla in the sea of Sicily, by reason of the huge and ragged rocks reaching into the sea, from which the waves rebounding with violence, make a great noise and roughness in the water."[2] Martyr later offers a comparable description of the Gulf of Paria lying between Trinidad and modern-day Venezuela, where "the waters are constrained together by the bending sides of that great land, and by the multitude of Islands lying against it, as the like is seen in the straits or narrow seas of Sicily where the violent course of the waters cause the dangerous places of *Scylla* and *Charybdis*."[3] In both passages, Martyr (from whom, given his humanist training, classical allusions might be expected) treats Homer's landmarks not as mythical wonders but as literal features of the landscape, in no way exceeding the cognate features of the New World. More interesting still is Martyr's admission in the first passage of the literary origins of his comparison: Scylla is a place "we read of," unexperienced outside the covers of a book.

This is an important distinction to draw about Renaissance geographical literature: it is both literal-minded and literary-minded in its use of sources. Martyr expects his allusions to be taken seriously as reasonable representations of the New World landscape, not as instances of hyperbole designed to return that landscape to the realm of the fabulous. For example, Martyr compares the Atlantic and Caribbean islands to the Nereids at several points in the *Decades*;[4] rather than proposing that these islands resemble transmogrified sea-nymphs, he invokes the time-honored association of the Nereids with the Aegean Islands. The attempt is to describe the Atlantic and the Caribbean in the familiar terms of the Mediterranean. At the same time, Martyr's references to the Nereids reflect limited experience of the Mediterranean and none at all of the New World—Martyr in fact did very little traveling—but instead display his considerable erudition. He attempts to describe a real world, but his descriptions are often drawn from another literary world where the question of "reality" does not necessarily arise.[5]

There are striking affinities between Martyr's description of the early Spanish voyages and Spenser's description of Guyon's voyage—a voyage with a literal quality that allegorizing critics frequently neglect to study. Though the various perils that Guyon encounters have bluntly allegorical titles and overtly mythological antecedents, this should not forestall an understanding of those perils as actual and physical within the limits of Spenser's narrative. Beyond their obvious significations, such titles and antecedents have a heuristic value not unlike Martyr's use of the Nereids to describe newly discovered islands; they lend form and color to a landscape that hovers on the very edge of the knowable.[6]

That Spenser has a literal voyage in mind is suggested by some of the perils he chooses to incorporate. After the sequence of named hazards ending with the Whirlpool of Decay in stanza 20, Guyon's boat runs into difficulties that present no clear-cut allegorical meaning. First, Guyon confronts "the great sea puft vp with proud disdaine" (12.21.7); this opens up a possible thematic connection with the figure of Disdain in canto 7, but the elaboration in stanza 22 indicates only that Guyon and his companions are battling a typical "angry" sea, the singular difference being that "not one puffe of wind there did appeare" (12.22.5).[7] The rough seas appear to be caused not by weather but by the approach of the sea monsters in the ensuing stanzas. These monsters are "dreadfull pourtraicts of deformitee" (12.23.5), and Spenser reiterates the idea by using "deforme" and "deformed" in close proximity (12.24.9, 25.2). Yet this emphasis on deformity has no obvious allegorical outcome; it is actually due to the fact that, as the Palmer explains, "these same Monsters are not these in deed, / But are into these fearefull shapes disguiz'd / By that same wicked witch" Acrasia (12.26.2–4). Acrasia's witchcraft may also account in part for the absence of wind on the rough sea.

The supernatural aspects of this seascape are indivisible from the natural; with or without wind a rough sea is still rough, and "disguiz'd" or not these sea monsters are close kindred to the creatures that Renaissance mariners regularly observed. The same consideration also applies to the descent of the "grosse fog" (12.34.5) and the eerie apparition of the birds in stanzas 35–37. The fog leads Spenser to remark, proverbially, that "Worse is the daunger hidden, then descride" (12.35.5)—a remark that could apply to dangers throughout *The Faerie Queene*. Yet fog is perhaps the most common hazard of navigation, generally encountered where cold water meets a warm landmass. Likewise, the flocking of "all the nation of vnfortunate / And fatall birds" (12.36.1–2), uncanny and ominous though it appears, is also a patent sign that Guyon's boat is approaching a landfall, which immediately appears at the end of stanza 37.

Whatever other significance this perilous voyage possesses, it is significant *as* a voyage. Guyon's hardships resemble those undergone by the various explorers in Martyr's text. Indeed, there are times when perils of the kind described in the *Decades* offer more immediate and vivid models than could be found in Spenser's traditional sources—as in, for example, the following detailed description of an inland "sea" occupying an island valley:

> It hath many swallowing gulfs, by the which, both the water of the sea springeth into it, and also such as fall into it from the mountains, are swallowed up. They think that the caves thereof, are so large and deep, that great fishes of the sea pass by the same into the lake. Among these fishes, there is one called *Tiberonus* which cutteth a man in sunder by the middest of one snap with his teeth, and devoureth him. . . . That salt lake is tossed with storms and tempests: And oftentimes drowneth small ships or fisher boats, and swalloweth them up with the mariners: In so much that it hath not been heard of, that any man drowned by shipwreck, ever plunged up again, or was cast on the shore, as commonly chanceth of the dead bodies of such as are drowned in the sea. These tempests, are the dainty banquets of the *Tiburones*.[8]

The passage conveys a sense of myths transcended, of marvels that overshadow their precursors because they are (or are believed to be) simple matters of fact. It is for this reason that such a text, given its availability, would be as open to Spenser as a source as the classical and Italianate sources usually cited for Book 2. More obviously, a description of a lake that mirrors the ocean in every respect except proportion would be quite relevant to *The Faerie Queene,* in which landscapes seem to exist on a scale more appropriate to the size of pages, stanzas, and printed words. Guyon's voyage across this dangerous sea is made, after all, in a "nimble boate" (12.38.2) rowed by a single boatman.

In the same way, Martyr's references to *tiburones* (i.e., sharks) and other dangerous sea creatures might provide the narrative grounds, if not the actual catalog of names, for the onslaught of monsters in stanzas 22–26.[9] Martyr recounts several tales of such creatures endangering Alonso de Ojeda's expeditions. In the first, the creature is a being of mysterious provenance, vaguely reminiscent of Grendel's mother: "a monster coming out of the sea, came upon one of them secretly and carried him away by the middest out of the sight of his fellows to whom he cried in vain for help until the beast leapt into the sea with her prey."[10] This creature appears to be kin to Spenser's "griesly Wasserman, that makes his game / The flying ships with swiftnesse to pursew" (12.24.3–4). In Martyr's second account, it is clearly a whale that sinks one of Hojeda's small craft: "Some of their fellows affirm that they plainly saw a fish of huge greatness swimming about the brigantine (for those seas bring forth great monsters) and that with a stroke of her tail, she broke the rudder of the ship in pieces: which failing, the brigantine being driven about by force of the tempest, was drowned."[11] It is possible that "sea-shouldring Whales" and "Mighty *Monoceroses,* with immeasured tayles" (12.23.6, 9) would have found their way into Spenser's catalog in any case. But Martyr offers a story that meets the circumstances of Spenser's narrative, in which "huge Sea monsters" (12.22.9) threaten a small vessel in the midst of a tempest, and Martyr's story is intended to be credible.

I hope that the discussion above offers strong and persuasive reasons for treating Eden's edition of Martyr as an important source for canto 12. (As far as I know, it has not been cited before as a direct influence on Spenser's narrative.) Yet to entertain the idea that the *Decades,* and perhaps other works like it, have affected the design of *The Faerie Queene* is to have to adjust one's sense of the allegory, historical or otherwise, in this initial episode of canto 12. The adjustment leads not so much away from the need for allegorical interpretation as toward an appreciation of the plausibility of certain features of Guyon's voyage. The reader is returned to what A. C. Hamilton once described as the imperative "to focus upon the literal level in its depth."[12] For example, Guyon encounters what seems to be an overtly symbolic obstacle in the Rock of Vile Reproach,

> On whose sharpe clifts the ribs of vessels broke,
> And shiuered ships, which had bene wrecked late,
> Yet stuck, with carkasses exanimate
> Of such, as hauing all their substance spent
> In wanton ioyes, and lustes intemperate,
> Did afterwards make shipwracke violent,
> Both of their life, and fame for euer fowly blent.
> (12.7.3–9)

This stanza comes across on first reading as an example of (so to speak) transparent allegory; the only wreckage here is the moral wreckage of intemperate mankind. In the next stanza, however, Spenser adds details that tilt the description toward a sense of actuality:

> A daungerous and detestable place,
> To which nor fish nor fowle did once approch,
> But yelling Meawes, with seagulles hoarse and bace,
> And Cormoyrants, with birds of rauenous race,
> Which still sate waiting on that wastfull clift.
> (12.8.2–6)

The presence of what are essentially very ordinary seabirds—their emblematic meaning notwithstanding—lends a quality of naturalism to the Rock, though this begins to dissipate upon the news that the birds are waiting "For spoyle of wretches, whose vnhappie cace, / After lost credite and consumed thrift, / At last them driuen hath to this despairefull drift" (12.8.7–9). The tilt back toward allegory seems to be confirmed by the Palmer's admonition in the following stanza:

> Behold th'ensamples in our sights,
> Of lustfull luxurie and thriftlesse wast:
> What now is left of miserable wights,
> Which spent their looser daies in lewd delights,
> But shame and sad reproch, here to be red,
> By these rent reliques, speaking their ill plights?
> Let all that liue, hereby be counselled,
> To shunne *Rocke of Reproch,* and it as death to dred.
> (9.2–9)

The prevailing impression is that Guyon has just successfully sailed past a rhetorical trope in a largely "tropical" sea.

A few stanzas later, however, this impression becomes less certain. The boat passes by the Quicksand of Unthriftyhead, where Guyon and his companions see

> a goodly ship . . .
> Laden from far with precious merchandize,
> And brauely furnished, as ship might bee,
> Which through great disauenture, or mesprize,
> Her selfe had runne into that hazardize;
> Whose mariners and merchants with much toyle,
> Labour'd in vaine, to haue recur'd their prize,
> And the rich wares to saue from pitteous spoyle,
> But neither toyle nor trauell might her backe recoyle.
> (12.19)

Stanza 19 presents an allegory of a peculiarly topical sort. Spenser describes the wreck of the treasure ship with a notable absence of metaphysical trappings; the stanza reads not as a caution against profligacy but as an account of such a ship going down as the result of "disaduenture, or mesprize" rather than the habit of intemperance. Interestingly enough, the vice here is a matter of economy, and one of fairly thin spiritual dimensions. Given that Spenser is writing at or shortly after the height of Philip II's naval power, when Philip's ships were renowned for coming "Laden from far with precious merchandize, / And brauely furnished, as ship[s] might bee," this argosy could readily be construed as hailing from Spain, and its catastrophe understood as exemplary less for mankind in general than for the English in particular. It is, moreover, an example based in fact; the English under privateers like Drake, John Hawkins, and Thomas Cavendish had participated in the wreck of such "goodly" Spanish ships—ships doomed because they were so unthriftily loaded with American goods.

This is not to say that the meaning of the example is clear at this point, or at any point, in canto 12; it comes to little more here than the necessity of rowing onward. But the presence of the ship in stanza 19 resonates with the images in the previous stanzas and leads toward the notion that the Palmer's statement "Behold th'ensamples in our sights, / Of lustfull luxurie and thriftlesse wast" is in many respects a literal statement with historical ramifications. Guyon voyages past contemporary events as well as allegorical images; he is a witness to "the ribs of vessels broke, / And shiuered ships, which had bene wrecked late" much as Drake, Hawkins, and Cavendish were witnesses to such things. The English reader might not encounter such things directly but could experience them vicariously by way of the voyage literature, full of instances of the "shipwracke violent" of figures like Nicuesa and Ojeda. The narratives in the *Decades* offer moral fables without much in the way of moral commentary; the epitome is in the event itself. This characteristic of the voyage literature has a place among the more familiar tropes of *The Faerie Queene,* for Spenser is concerned with the immediate condition of England. It is appropriate that he should inform the poem with a quality of historical, political, and geographical immediacy. This sense of immediacy follows Guyon throughout his voyage, becoming still more powerful when the little boat at last reaches the Bower of Bliss.

My account has now arrived at the locus around which the recent case for Spenser as a poetic interpreter of Elizabethan colonialism has been developed.[13] The case has been built not only upon the physical features of the Bower of Bliss but also upon Guyon's actions within it. These actions compose the most forceful, as well as the most controversial, statement of Guyon's identity in the entire book. Both the force and the controversy emerge from Guyon's destructiveness, his capacity for making "of the fairest late, now . . . the fowlest place" (12.83.9). The deeds in stanza 83 seem antithetical to what the reader expects from a knight whose greatest virtue is temperance. There is a critical tradition, almost certainly originating with Lewis's observations on Spenser in *The Allegory of Love,* that Guyon's "rigour pittilesse" (12.83.2) is justified by the innate viciousness of Acrasia and her domain, a viciousness that Guyon must undo in order to return the world of the poem to equilibrium.[14] Yet this version of events has not ultimately proven satisfactory, for aesthetic and perhaps logical reasons. After years of critical examination, stanza 83 still retains its power to jar; it departs abruptly from the tone Spenser has set for much of Book 2, and it does not fit comfortably with the emphasis on Guyon's resistance to action throughout canto 12. There are mild anticipations of stanza 83 in Guyon's behavior earlier in the canto, as when he upends Genius's winebowl and breaks his staff (12.49.8–9), and when he shatters the cup of Excess (12.57.3–5). But these acts are not in the same category as felling, defacing, spoiling, suppressing, burning, and razing. Stanza 83 is also oddly discrete; its mayhem extends to neither the preceding nor the following stanza.

Part of the problem in interpreting the conclusion of Book 2 comes from the natural tendency of readers to equate the culmination of a narrative with the culmination of a particular theme or line of character development. The impulse is to view the last handful of stanzas in canto 12 as somehow putting Guyon's character to rest. This is where the difficulty lies: if Spenser has nothing more to say about Guyon, then the contradictions suggested by stanza 83 seem much more glaring. In fact, Spenser is not entirely through with Guyon. Rather than returning to Gloriana's court, Guyon and Arthur continue at the beginning of Book 3 "To hunt for glorie and renowmed praise; / Full many Countries they did ouerronne, / From the vprising to the setting Sunne, / And many hard aduentures did atchieue." (3.1.3.3–6). Evidently, these two have been traveling ever farther from their respective homelands. After passing "Through countries waste, and eke well edifyde," they once more enter a wilderness, a "forrest wyde" inhabited only by "Beares, Lions, and Buls" (3.1.14.2, 5, 9). After Guyon's richly ambiguous encounter with Britomart—where the male figure of Temperance is unhorsed by the female figure of Chastity—he and Arthur ride off in pursuit of Florimell (3.1.18). Guyon appears again in canto 4, still chasing after Florimell; confronted with a fork in the path, Guyon and Arthur agree to "dispart" themselves (4.46.8), and again Guyon rides out of the narrative. Clearly, Spenser does not intend that his readers think of Guyon's story as over; Guyon, like so many other figures in *The Faerie Queene,* is still "in process" even when the substantive focus of Spenser's narrative shifts elsewhere. Indeed, Guyon returns for another cameo appearance in Book 5, when he at last catches up with Braggadocchio and reclaims his stolen horse (5.3.29–36).

This leads back to the problem of stanza 83. To regard that stanza as

Guyon's terminus is to neglect the open-ended quality of Spenser's narrative technique, a quality reflected not only in plot and character but also in theme and structure. *The Faerie Queene* constantly holds out the prospect and promise of closure, yet the 1596 edition of the poem ends with an equivocal fragment that implies, depending on one's interpretation, either finality or continuity.[15] Whether this lack of closure results from Spenser's own aesthetic inclinations or from his position at a curious intellectual and political watershed in English history, one distinguished by a surplus of speculation and a dearth of definite conclusions, is certainly subject to conjecture. However its contexts are understood, *The Faerie Queene*'s poetic openness still confronts contemporary readers with the need to make frequent adjustments in their understanding and expectations of the text.

Instead of regarding Guyon's actions at the end of canto 12 as the final stage in Spenser's "definition" of Guyon, readers have to assume a more tentative stance, viewing those actions as a further variation on a still-provisional definition, not unlike the variations that Spenser introduces during Guyon's encounter with Mammon. Guyon may be more "knowable" after canto 12, just as he was after canto 7, but readers do not know everything about him. If the ideas personified in Guyon are supposed to extend into moral and political praxis in a real world, then Spenser has left more than ample space at the end of Book 2 for such an extension, since his literary enterprise is speculative rather than predictive. A reasonably accurate prediction of a particular outcome—for instance, the eventual English annexation of North America—would require a fund of knowledge large and definitive enough to justify making the prediction in the first place; a false prediction—false because based on inadequate knowledge—is a worthless one. Given the failure of the English settlement of America during Spenser's lifetime, he was unlikely to present any sort of confident imperial "program," poetic or otherwise.

Guyon's conduct in the Bower of Bliss, then, represents a further, though not a complete, understanding of the relationships that Spenser has been considering throughout Book 2. Acrasia, in this account, represents Spenser's notion of a "native" culture, or at least of certain aspects of such a culture. Guyon's destruction of her bower does not translate readily into Spenser's advocacy of enslavement or suppression of natives à la *A View*, but it does suggest that his judgment of native culture is cautionary at best. I will also argue that his judgment of Old World culture (represented mainly in the figure of Verdant) is cautionary in the same fashion. The remainder of this chapter, then, will be devoted to the peculiarly complex historical dimension of the Bower of Bliss, and to Spenser's efforts to close the book on the act of exploration which has engaged him throughout Book 2.

Along with its obvious connections with Book 16 of Tasso's *Gerusalemme Liberata* in particular, the Bower resembles landscapes traversed in the narratives of Martyr, Gómara, Zárate, and Oviédo. A number of the resemblances are superficial; some are more basic and bear significantly on Spenser's peculiar vision of an English adventurer in a new world. Both Cain and Bernhart draw the analogy between Spenser's grapes "of burnisht gold, / So made by art, to beautifie the rest" (12.55.1–2), as well as the "trayle of yuie" made "of purest gold" (12.61.2, 1), and contemporary descriptions of artificial gardens alleged to exist in various places in the Indies.[16] Zárate describes one such garden near the island of Puná in the Gulf of Guayaquil off the Peruvian coast: "In another little Island adjoining to the same, they found a house and a garden plot or orchard within the same, having little trees and plants therein, made of silver and gold."[17] Yet this is only one of numerous "boweresque" images to be found in the voyage literature, especially in Martyr's *Decades.*

Spenser depicts the interior of the Bower as "A large and spacious plaine, on euery side / Strowed with pleasauns, whose faire grassy ground / Mantled with greene, and goodly beautifide / With all the ornaments of *Floraes* pride" (12.50.2–5). These ornaments benefit from the Bower's subtropical climate:

> Thereto the Heauens alwayes Iouiall,
> Lookt on the louely, still in stedfast state,
> Ne suffred storme nor frost on them to fall,
> Their tender buds or leaues to violate,
> Nor scorching heat, nor cold intemperate
> T'afflict the creatures, which therein did dwell,
> But the milde aire with season moderate
> Gently attempred, and disposd so well,
> That still it breathed forth sweet spirit and holesome smell.
> (12.51)

This appears to be a variation on the way in which Tasso introduces Armida's garden in *Gerusalemme Liberata.*[18] At the same time, Spenser's stanza presents a scene not very different from what Columbus's crew saw on the third voyage, suggesting that the Bower's link to Tasso's mid-Atlantic paradise is only part of a more complicated provenance: "looking out of their ships, they might well perceive that the Region was inhabited and well cultured. For they saw very fair gardens, and pleasant meadows: from the trees and herbs whereof, when the morning dews began to rise, there proceeded many sweet savors."[19] Later in the *Decades* Martyr describes the remarkable climate of Trinidad: "In . . . places, the Island enjoyeth perpetual spring time, and is fortunate with continual summer and harvest. The trees flourish there all the whole year: And the meadows continue alway green. All things are exceeding fortunate, and grow to great perfection."[20]

In addition to the physical resemblances between the Bower and Martyr's Caribbean landscapes, there are similarities between occurrences in canto 12 and events recounted in the *Decades*. Guyon and the Palmer are twice offered wine on their trek, first from Genius's "mighty Mazer bowle" (12.49.3) and later from Excess's "Cup of gold"; in both instances Guyon overturns the offering. The second instance is the most interesting, because Spenser shows Excess actually making the wine:

> In her left hand a Cup of gold she held,
> And with her right the riper fruit did reach,
> Whole sappy liquor, that with fulnesse sweld,
> Into her cup she scruzd, with daintie breach
> Of her fine fingers, without fowle empeach,
> That so faire wine-presse made the wine more sweet:
> Thereof she vsd to giue to drinke to each,
> Whom passing by she happened to meet:
> It was her guise, all Straungers goodly so to greet.
> (12.56)

This drink, instantly fermented between the finger and the cup, is not wine under the traditional European definition. Evidently the "liquor" is in the fruit itself, which Spenser does not identify as a grape and which is already full of "lushious wine" (12.54.4) while still on the vine. It is wine befitting a place where "some [trees] were full of blossoms and flowers, and other laden with fruits."[21] It is also wine made to be offered to "Straungers," with the double sense of "unknown persons" and "foreigners." In the *Decades,* several participants in Columbus's third voyage attend a banquet hosted by the native nobility: "After that our men, and their Princes were set, their waiting men came in laden, some with sundry delicate dishes, and some with wine. But their meat, was only fruits: those of diverse kinds and utterly unknown to us. Their wine was both white and red: not made of grapes, but of the liquor of diverse fruits, and very pleasant in drinking."[22] The uncultivated beverage that Excess offers Guyon seems closer to this "liquor of diverse fruits" than to conventional wine; in refusing it, Guyon turns away not only from "excess" but also from an unfamiliar, possibly threatening novelty.

A more striking parallel emerges from Martyr's account of Columbus's second voyage in the manner in which the admiral's party is entertained at the court of the chieftain Anacachoa:

> there met them a company of xxx. women, being all the kings wives and concubines, bearing in their hands branches of date trees, singing and dancing: They were all naked, saving that their privy parts were covered with breeches of gossampine cotton. But the virgins, having their hair hanging down about

> their shoulders, tied about the forehead with a fillet, were utterly naked. They affirm that their faces, breasts, paps, hands, and other parts of their bodies, were exceeding smooth, and well proportioned: but somewhat inclining to a lovely brown. They supposed that they had seen those most beautiful *Dryads*, or the native nymphs or fairies of the fountains whereof the antiquities speak so much.[23]

This produces some rather unexpected, even startling echoes with Spenser's description in stanzas 63–68 of Guyon's meeting with the two maidens frolicking in the fountain—echoes heard despite the fact that the episode derives quite closely from Book 15 of *Gerusalemme Liberata*, in which Charles and Ubaldo encounter two very similar maidens. There is also the fact that Edward Fairfax relied heavily on material in canto 12 to flesh out his 1600 translation of Tasso's poem.[24] Yet the existence of this intertextual network does not completely resolve the question of Spenser's influences in canto 12. Tasso, too, had an interest in the new discoveries, as evidenced in the prophetic tribute to Columbus delivered to Charles and Ubaldo by their splendid fairy-like pilot on the journey to Armida's island:

> A knight of Genes shall have the hardiment
> Upon this wond'rous voyage first to wend;
> Nor winds nor waves that ships in sunder rent,
> Nor seas unus'd, strange clime, or pool unken'd,
> Nor other peril nor astonishment,
> That makes frail hearts of men to bow and bend,
> Within Abila's strait shall keep and hold
> The noble spirit of this sailor bold:
>
> Thy ship, Columbus, shall her canvas win
> Spread o'er that world that yet concealed lies;
> That scant swift Fame her looks shall after brin,
> Though thousand plumes she have and thousand eyes:
> Let her of Bacchus and Alcides sing,
> Of thee to future age let this suffice,
> That of thine acts she some forewarning give,
> Which shall in verse and noble story live.
>
> (15.31–32)[25]

Fairfax has translated Tasso's "istoria" as "story," which partially obscures the meaning of the last line; Tasso must have been thinking here of an actual historical work, and for him the best account of Columbus's exploits would have been the same as for Spenser—Martyr's *Decades*. The presence of Columbus in *Gerusalemme Liberata* serves to suggest that the passage about Anacachoa's maidens, and other such passages out of Martyr, cannot be easily dismissed as sources for canto 12, even when more obvious sources exist. Guyon's encounter with the maidens at the fountain conveys a sense of literary "refraction," of the layering of source upon source, reference upon reference. In displaying his obligation to Tasso, Spenser implies Tasso's obligation to *his* sources, and Martyr is thus drawn into a thematic triangulation with *Gerusalemme Liberata* and *The Faerie Queene;* Spenser and Tasso end up rowing in the same direction, toward a vanishing point on the Atlantic horizon.

As interesting as such analogies are, they only carry the reader part way toward an understanding of Spenser's purposes in devising the Bower episode. Further study of the voyage literature uncovers passages and ideas that help to limn more sharply the inner workings of the canto and to make better sense of its culminating stanzas. Guyon's encounter with the maidens at the fountain provides an appropriate point of departure. While gazing on these provocative apparitions, Guyon begins "secret pleasaunce to embrace" (12.65.9). Greenblatt uses the passage to call attention to both the attractiveness of this vision of natural sexual license and the necessity on the part of "civilized" man to repudiate it, for "to succumb to that beauty is to lose the shape of manhood and be transformed into a beast."[26] This necessary repudiation of what is perceived as the amorality of native cultures occurs quite transparently in the voyage literature. Eden's translation of extracts from Sebastian Munster's *Cosmographia Universalis* (portions of which Munster borrowed from the *Decades*) includes a description of the Indian lifestyle that characteristically mingles fascination—amounting almost to admiration—with disgust:

> Their bodies are very smooth and clean by reason of their often washing. They are in other things filthy and without shame. They use no lawful conjunction of marriage, but every one hath as many women as him listeth, and leaveth them again at his pleasure[.] The women are very fruitful, and refuse no labor all the while they are with child. They travail in manner without pain, so that the next day they are cheerful and able to walk. Neither have they their bellies wrimpled, or loose, and hanging paps, by reason of bearing many children.[27]

Along with demonstrating a metonymic relationship to the continuously "fruitful" landscape they inhabit, Munster's Indian women fulfill one of the primal fantasies of Western culture, one that has lingered into the twentieth century: the fantasy of childbirth without suffering; or, to place the fantasy in another light, of sex without consequences.[28] These women look and behave no differently before or after childbirth. Moreover, neither the men nor the women display any sense of shame, a sense which is so often dependent on an acknowledgment of the potential effects of "shameful" activity: pain, dislocation, physical transformation. A society that exists without pain or shame is one that, from the Old World perspective at least, most closely resembles the "society" of the animals. The fear of bestiality, of the human being regressing fully to the state of an unreasoning animal, is as primal to Western societies as the fantasy of painless childbirth.

In the voyage literature, bestiality is closely linked, naturally enough, with cannibalism; for the sixteenth-century traveler and scholar there is little difference between natives understood as indiscriminate carnivores ("They found also in their kitchens, mans flesh, ducks flesh, and goose flesh, all in one pot."[29]) and natives regarded as mere predatory animals. This de facto equivalence leads to one of the more peculiar passages in the *Decades*. In order to evoke the grisly strangeness of the cannibals' behavior, Martyr falls back on analogies to the eating customs of civilized Europeans:

> Such children as they take, they geld to make them fat as we do cock chickens and young hogs, and eat them when they are well fed: of such as they eat, they first eat the entrails and extreme parts, as hands feet, arms, neck, and head. The other most fleshy parts, they powder for store, as we do pestles of pork and gammons of bacon. Yet do they abstain from eating of women and count it vile. Therefore such young women as they take, they keep for increase, as we do hens to lay eggs.[30]

A curious feature of this passage is that Martyr never compares the cannibals themselves to barnyard animals, only their victims. The cannibals' feat of domestication provides for both food and progeny, and in both of these particulars remains uncomfortably close to the domestication practiced in the Old World, not only in its superficial features but in the fact that the onus of "beastliness" comes to reside with the domesticated, not with the domesticator. Thus Martyr offers an instance of (probably unintended) cultural relativism: the cannibals, from their own limited perspective on the world, are "farmers" and "herdsmen," and the culture they enact is not wholly invalidated in Martyr's rendering of it. Why should this be so? A possible answer is that the cannibals' "animals," who would seem at first to resemble the cannibals both physically and culturally, are finally too stupid or weak to be treated as anything but animals. Behind Martyr's evident repulsion at cannibalism there emerges a certain empathy with the cannibals' position in the pre-Columbian hierarchy of the New World. The fear of bestiality on display here manifests itself as a fear not of uncontrolled aggression, but of uncontrolled passivity.

It is not surprising, then, to find considerable anxiety expressed in the *Decades* over the appearance of passivity in a sexual guise. Martyr's report that the cannibals geld children "as we do cock chickens and young hogs" is only one indication of a concern about the threat of emasculation and the appearance of effeminacy in the New World. This issue occupies much of Greenblatt's attention in his discussion of the Bower and prompts his only quote from the *Decades:* "Of the sweet savors of these lands, many things might be spoken, the which because they make rather to th[']effeminating of the minds of men, then for any necessary purpose, I have thought best to omit them."[31] This is perhaps the most elliptical example of Martyr's concern with the dangers of effeminacy. Earlier, he has made clear what he means by "sweet savors"—namely, perfumes: "the most part of the Spaniards do laugh them to scorn which use to wear many stones . . . Judging it to be an effeminate thing, and more meet for women then men. . . . They think it no less effeminate for men to smell of the sweet savors of Araby: And judge him to be infected with some kind of filthy lechery, in whom they smell the savor of musk or *Castoreum*."[32] Earlier still, Martyr cites an event that pointedly illustrates the Spaniards' intolerance of perceived effeminacy in appearance or behavior. In his account of Balboa's explorations in Darien, he describes Balboa's discovery of apparent berdache practices in the village of Quaraca: "*Vascus* found the house of this king infected with most abominable and unnatural lechery. For he found the kings brother and many other young men in womens apparel, smooth and effeminately decked, which by the report of such as dwelt about him, he abused with preposterous venus. Of these about the number of forty, he commanded to be given as prey to the dogs."[33] Here Martyr presents an example of "rigour pittilesse" fully as terrible as Guyon's in stanza 83—more terrible, indeed, because it claims the authenticity of a historical event in which human lives were lost.[34]

The perspective represented above enters canto 12 in some fairly obvious ways. There are hints of pederasty within the Bower, as in the fountain's depiction of "naked boyes . . . playing their wanton toyes" (12.60.6, 8) and in the "lasciuious boyes, / That euer mixt their song with light licentious toyes" around Acrasia's bed (12.72.8–9). Most obviously there is the figure of Genius,

> A comely personage of stature tall,
> And semblaunce pleasing, more then naturall,
> That trauellers to him seemd to entize;
> His looser garment to the ground did fall,
> And flew about his heeles in wanton wize,
> Not fit for speedy pace, or manly exercize.
> (12.46.4–9)

Given the tension between "natural" and "artificial" throughout canto 12 and the metaphorical examples of that tension in the text (golden fruits among the real, etc.), it does not seem implausible to regard Genius's "semblaunce pleasing, more than naturall" as the result of, among other things, make-up.[35] A few stanzas later, Spenser indicates that Genius is "With diuerse flowres . . . daintily . . . deckt" (12.49.1). This floral display becomes important at the canto's conclusion, where Acrasia's captive Verdant is described as also, in a sense, being "deckt" with flowers. He sleeps with Acrasia "Vpon a bed of Roses" (12.77.1); a song is sung over him in which a rose appears as the central metaphor (12.74–75); and his face, where signs of "manly sternnesse" can still be seen (12.79.6), appears to be growing flowers rather than a beard: "on his tender lips the downy heare / Did now but freshly spring, and silken blossomes beare" (12.79.8–9). Such flowers would, of course, be likely to give off "sweet savors."

Among these examples, the crucial one is offered by Verdant, whose presence in the Bower leads back to the problem Spenser has been weighing so carefully throughout Book 2. The traditional view of Guyon's test in the Bower is that it occurs largely in relation to Acrasia and her various minions. I would argue, however, that the innermost conflict in canto 12 is actually between Guyon and Verdant—here understood as the final version of Spenser's antithetical knight—and that this conflict represents yet another confrontation between English identity and what Spenser considers the most important "global" threat to that identity.

While Genius's effeminacy is a given, an unexceptional feature of a realm controlled by an arch-feminine enchantress, Verdant's is not. He has had to surrender willfully to his effeminate state, and Spenser leaves no doubt that Verdant is responsible for his own condition: "certes it great pittie was to see / *Him* his nobilitie so foule deface" (12.79.3–4; my emphasis). This willfulness allies him with the notorious Grill, formerly of "hoggish forme" and still of "hoggish mind" (12.86.9, 12.87.8), who "chooseth, with vile difference, / To be a beast, and lacke intelligence" (12.87.4–5).[36] The phrase "vile difference" implies an element of sexual perversity in Grill's generally beastly demeanor. Verdant also belongs among the company of Acrasia's other victims, the "wild-beasts" (12.84.5). They, too, are victims only in a qualified way. By transforming them from men into beasts, Acrasia has merely made visible their inward weaknesses; they are "turned into figures hideous, / According to their mindes like monstruous" (12.85.4–5). When the Palmer returns them to their former shapes,[37] these men look "vnmanly" (12.86.3)—an epithet which, in light of Spenser's earlier allusions to effeminacy, carries considerable weight. Finally, Verdant echoes Cymocles during *his* sojourn in the Bower in canto 5. Summoning Cymocles to aid his brother, Atin found him "Hauing his warlike weapons cast behind" and "Mingled emongst loose Ladies and lasciuious boyes" (5.28.7, 9), and berated him as "*Cymocles* shade, / In which that manly person late did fade. . . . Vp, vp, thou womanish weake knight" (5.35.4–5, 36.2). The speech shows Atin's characteristically abrasive mode, yet it effectively captures Cymocles's predicament, transformed like Verdant into a willfully "womanish" inhabitant of the Bower.

At the same time, canto 12 is liberally scattered with hints that Verdant fills the role of the conquistador. When Guyon and the Palmer pass through Genius's gate they see portrayed there "all the famous history / Of *Iason* and *Medea*" (12.44.3–4)—an interesting choice of myth on Spenser's part, for Jason's story involves an ocean voyage in search of a golden object, a voyage which ultimately ends in tragedy and degradation owing to Jason's "falsed faith, and loue too lightly flit" (12.44.7). The story also involves the bloody sacrifice of an innocent, Medea's brother, pictured in the mural as "a piteous spectacle" (12.45.7). The entire mural is "with gold besprinkeled" (12.45.8). The fountain in which Guyon sees the maidens is made "Of richest substaunce, that on earth might bee" (12.60.2) and strangely resembles a ship floating in a pool; it "seemd the fountaine in that sea did sayle vpright" (12.62.9). Verdant himself possesses a shield "full of old moniments," now "fowly ra'st, that none the signes might see" (12.80.3–4). The word "moniments" possesses obviously multiple senses, suggesting not only the shield's insignia but also the commemoration of great achievements, and even the depiction of commemorative structures.

Verdant has succeeded in tearing these "moniments" down: "Ne for them, ne for honour cared hee, / Ne ought, that did to his aduancement tend, / But in lewd loues, and wastfull luxuree, / His dayes, his goods, his bodie he did spend" (12.80.5–8). The passage implies that Verdant's decline is more than moral; it is political and economic as well. "Aduauncement" in the sixteenth century is a word heavily freighted with the ethos of a court society dependent on royal patronage and fueled by an often entirely pragmatic sense of "honour." And Spenser studiously inserts "goods" before "bodie" in his list of Verdant's losses.

Verdant, like his ill-fated predecessors Mordant and Cymocles, is a poor steward of the "goods" of this new world. He reflects a growing sense (much of it emanating from the Spanish chronicles themselves) that the Spaniards had not made good use of their great windfall, that they had instead embraced the most repellent characteristics of the natives and made them their own—making themselves worse in the process, since the Spaniards could not plead ignorance as the Indians could of a better, truer way of life. Evidence of trepidation over the possibility of Spanish depravity in the New World can be found in Martyr's comment on the wearing of "stones" and "the sweet savors of Araby." When Martyr says "the Spaniards, do laugh them to scorn" who wear these things, he is using the pronoun in a general way, without specific reference to the Indians. The condemnation of the use of jewelry and perfumes extends to the Spaniards as well, for Martyr states that "the most part," not all, of the Spaniards laugh at such fashions.

This is a relatively subtle example. Eden's translation of the *Decades* is replete with marginal glosses referring to "Liberty and Idleness" or "Idleness and Play,"[38] a number of which refer to the Indians (as in Martyr's remark that "The people of this Region are given to Idleness and play"[39]) but many of which actually refer to the Spaniards. For Eden as for Martyr, idleness and liberty both connote the sort of brutal sexual behavior displayed by Columbus's crews on Hispaniola:

> that kind of men (the Spaniards I mean which followed the Admiral in that navigation,) was for the most part unruly, regarding nothing but Idleness, play and liberty: And would by no means abstain from injuries: Ravishing the women of the Islands before the faces of their husbands, fathers, and brethren: By which their abominable misdemeanor, they disquieted the minds of all th[']inhabitants.[40]

This kind of behavior leads swiftly into social and political depravity as well:

> the Spaniards which he took with him into these regions, were given rather to sleep, play, and idleness, then to labor: And were more studious of sedition and news, than desirous of peace and quietness: Also that being given to licentiousness, they rebelled and forsook him, finding matter of false accusations against him, because he went about to repress their outrageousness.[41]

There is probably little need to note that the Idle Lake is a central feature of the landscape in Book 2, that a figure of mirth (i.e., play) controls its shores, and that, while Guyon traverses the lake successfully, its waters are closely linked with the bad ends of Pyrocles and Cymocles, both of them exemplars of "Liberty and Idleness."

The two passages above reflect the notion that sexual excess has a disruptive effect on the colonial polity and, by implication, on the state as a whole. For Martyr, however, the danger of sexual release is less daunting than the danger of political release; the latter is most likely to throw the colonial enterprise into disgrace. "Goods"—the goods of government, commerce, and military power—are more crucial than the body to the ordering, or disordering, of an empire.

It is against this background that Guyon's actions at the end of the canto should be judged. Though the Indians certainly bear frequent charges of idleness and sexual depravity in the voyage literature, it is clear that the Spaniards carry the onus almost as frequently. In at least one instance, from Zárate, the Indians become the accusers rather than the accused. Diego Almagro, one of Pizarro's chief accomplices and later one of his chief rivals in Peru, repeatedly engages with hostile natives during his initial journey down the west coast of South America:

> In this sort they [Almagro's men] went rowing with their *Canoas* against the current of the Sea, which always runneth Northward, and their way was Southward: and in this Navigation all along the Coast, the Indians assaulted them according to the custom of their Wars, thundering out cries and noise, calling them banished men, with hair on their faces, yea, such as were bred of the scum of the Sea, without any other Origin or Lineage, because the Sea had brought them thither: demanding

> also why they went like Vagabonds wandering the World: it should appear said they, that you are idle persons, and have not wherein to employ your selves, because you abide in no place, to labor and till the ground.[42]

Zárate's willingness to record the Indians' (apparently translatable) insults, when mere mention of the attack would suffice for narrative purposes, suggests that on some level he accepts the seriousness of the charge of vagabondage, even if it comes from an unlikely source. The account displays a nagging awareness among the Spanish that the nation had not devoted its best and brightest to the task of empire, that many of the rank and file among the conquistadors—and even some of the leaders (Almagro, for example)—could seriously be viewed as descended from "the scum of the Sea." The New World was full of cagey but not necessarily intelligent drifters who rowed in many respects "against the current."[43] The constant mingling of such men with alien cultures often led to a monstrous hybrid of Old World and New World vices.

From this perspective, Verdant's slumber in the Bower represents less his victimization by Acrasia than his collusion with her. Thus, it is appropriate that Guyon and the Palmer do not initially discriminate between the targets of their ambush in stanzas 81–82 of canto 12. Acrasia seeks to free herself from the Palmer's "subtile net" (12.81.4), but Spenser hastens to add in the first line of stanza 82, "And eke her louer stroue." Escape is as important to Verdant as it is to Acrasia. Spenser describes the two lovers as at first enduring an equivalent subjection: Guyon and the Palmer "tooke them both, and both them strongly bound / In captiue bandes, which there they readie found" (12.82.4–5). The reiteration of "both" in such a short span would appear to indicate Spenser's desire to place Verdant and Acrasia under roughly the same moral scrutiny.[44] The fact that the "captiue bandes" are already present in the Bower—whatever its overt allegorical significance—does not alter the point; since these "bandes" happen to be empty when Guyon and the Palmer approach, there is no way of knowing whether they are part of Acrasia's equipage or Verdant's.[45]

It is true that Verdant and Acrasia depart the Bower enduring very different sorts of captivity: Acrasia is bound "in chaines of adamant" (12.82.6), while Verdant is "soone vntyde" and made captive, so to speak, to "counsell sage in steed" (12.82.8, 9). This difference may be less problematical than it seems. Whatever allegorical construction is placed upon Acrasia's character, these "chaines of adamant" still intimate that she represents something larger and more elusive than Verdant does. Verdant, though, is not alien to Guyon's world; he is integral to that world, a knight despite his "warlike armes" having become "the idle instruments / Of sleeping praise" (12.80.1–2). As a familiar figure in the "community" of *The Faerie Queene,* Verdant is accessible, and capable of rehabilitation, in a way that Acrasia is not. Yet if the conclusion of the canto is viewed as an example of the Englishman literally teaching the Spaniard a lesson, Spenser is not at pains to declare what that lesson—Guyon's "counsell sage"—includes. At the end of Book 2, Verdant still stands uncorrected; he and Acrasia leave the Bower "sorrowfull and sad" (12.84.2), presumably less at the view of their moral depravity (*pace* Adam and Eve) than at the prospect of having to leave at all. By the beginning of Book 3, Verdant has already disappeared from Guyon's entourage.

The most likely possibility here is that Spenser intends Guyon himself to be the "lesson," an exemplar against which Verdant can judge his own failings as a knight and a crusader in foreign lands. This leads back to the question of the example that Guyon sets for the purposes of the canto. Clearly, Spenser wants to establish a degree of likeness between Guyon and Verdant, even while his main purpose is to distinguish them from one another. Without some likeness there would be little evidence that Guyon is being tested and that he is succeeding in the ordeal at the same points where Verdant succumbed. Guyon's ordeal mainly involves his much remarked dalliance with the maidens in the fountain: he flirts with the crime of idleness when, attracted to the spectacle of these "naked Damzelles" at play (12.63.6), he decides "to slacke his pace, / Them to behold" (12.68.4–5). The "secret signes of kindled lust" that cross his face (12.68.6) are kin to the suspicious concealments of Verdant's shield, where "none the signes might see." Indeed, Guyon almost falls into a kind of sensory vagabondage when his eyes begin "wandring" (12.69.2). His brush with a fate like Verdant's is quickly over, however, when Guyon is "rebukt" by the Palmer and "counseld well, him forward thence" to "draw" (12.69.2, 3). The Palmer's counsel seems to consist largely of offering his correct sense of his and Guyon's position, in both a narrative and a geographical sense: "Now, Sir, well auise; / For here the end of all our trauell is: / Here wonnes *Acrasia,* whom we must surprise, / Else she will slip away, and all our drift despise" (12.69.6–9). It is the Palmer, not Guyon, who best understands the requisites of this quest; Guyon's appropriate response is one of acquiescence. By orienting himself according to the Palmer's advice, he passes unscathed through the perils of the Bower.

Guyon's dependence on the Palmer in canto 12 may suggest that Spenser is proposing a model hierarchy of clerical and secular forces to be deployed when the English actually become possessors of new lands; the clerical presence guides, sanctions, and validates—gives identity to—the efforts of the secular actor. This beneficial relationship stands in contrast to the failure of a similar association in canto 1, when Amavia "in Palmers weed" seeks to rescue Mordant from the Bower but cannot save him (or herself) from the tragic consequences of his tryst with Acrasia. In chapter 2, I argued that Guyon and the Palmer represent reformed versions of the archetypes of knight and pilgrim characterized initially by Mordant and

Amavia. Spenser's "reformation" may be based on an awareness of the schism that had long since developed between secular and ecclesiastical authorities in the New World. The missionary orders commanded a great deal of respect within the Spanish sphere of influence, but too often the conquistadors—frequently men of little religious sensibility to begin with—disregarded the intentions of the friars who accompanied them.[46] Paradoxically, a voice as influential and articulate as Las Casas's only served to point out the friars' lack of much direct control over the chaotic behaviors of their countrymen.

Canto 12, by contrast, presents a commissioned soldier of the Fairy Queen relying almost completely on his clerical, or at least quasi-clerical, guide. The Palmer is generally allegorized as the figure of Guyon's reason, which Guyon alternately loses and regains in the course of the narrative. Yet it is worth asking if the Palmer's relationship with Guyon might not reflect Spenser's notion of the virtues of England's more "reasonable" religious establishment and its potential benefits to the project of overseas expansion. The English would, after all, not be sending priests and friars to the New World, but ministers of the reformed church.

Even so, this argument can be pressed too far. The concluding stanza of canto 12 probably forestalls a definitive statement, certainly any historically grounded one, about the Palmer and his relationship to Guyon. His last counsel to Guyon, forming the sentence with which Spenser chooses to end Book 2, does not exactly provide a ringing coda:

> Said *Guyon,* See the mind of beastly man,
> That hath so soone forgot the excellence
> Of his creation, when he life began,
> That now he chooseth, with vile difference,
> To be a beast, and lack intelligence.
> To whom the Palmer thus, The donghill kind
> Delights in filth and foule incontinence:
> Let *Grill* be *Grill*, and haue his hoggish mind,
> But let vs hence depart, whilest wether serues and wind.
> (12.87)

The Palmer's words indicate that the destruction of the Bower is incomplete, and will remain so. The Bower's depopulation does not extend to Grill; the disjunction of "Let *Grill* . . . But let vs" implies that Grill will remain behind, still pursuing "filth and foule incontinence." In its elliptical way, the whole stanza sounds distinctly similar to a much longer admonition from Eden's preface to the 1555 edition of the *Decades,* the preface that Willes omits in 1577:

> sure I am, that like as the slow and brutish wits, for the slenderness of their capacity and effeminate hearts, do never or seldom

> lift up their minds to the contemplation of gods works and majesty of nature, but like brute beasts looking ever downward, think the world to be in manner no bigger than their own dunghills and cages, little passing whether the Christian faith do spread through the world, or be driven to one corner: Even so all good wits and honest natures (I doubt not) will not only rejoice to see the kingdom of God to be so far enlarged upon the face of the earth, to the confusion of the devil and the Turkish Antichrist, but also do the uttermost of their power to further the same.[47]

What distinguishes Eden's scorning of the dunghill from the Palmer's is that Eden is proposing the institution of a grand quest, while the Palmer is proposing the abandonment of one. This is an extraordinary statement for the Palmer to make, especially given that it is his final statement. In effect he removes the impetus for continuing Guyon's intervention, an intervention that has been more decisive here than during any other episode in Book 2; he advises Guyon to turn away while he has the chance to do so, "whilest wether serues and wind."[48]

This could be interpreted as the counsel of resignation: something can be done about the noble Verdant but not about the savage Grill (though presumably Grill once belonged to the same class as Verdant). More importantly, the Palmer's final words serve indirectly to suggest that Guyon's destruction of the Bower in stanza 83 lacks many of the moral and metaphysical implications so often attributed to it. Guyon's "rigour pittilesse" may well represent neither his violent release after the repression of his sexual desires (see Greenblatt), nor a surplus of intemperate rage, nor Guyon's just revenge against an extreme perversion of the natural order, nor the high-handed advances of an anointed imperialist. If any of these constituted the primary motive behind Guyon's act, the Palmer could be expected to moralize upon the outcome in some way; in fact he has nothing whatsoever to say about the events in stanza 83. The ultimate incompleteness of the act—for Grill's continuing presence prevents the *tabula* from being utterly *rasa*—suggests instead that "the tempest of [Guyon's] wrathfulnesse" is directed toward a rather simple military, as opposed to militant, end. Together with the dispossession of its primary "settlers," Guyon's sacking of the Bower meets the useful, if limited, tactical purpose of preventing the same site from being resettled quickly (by either natives or colonists) and becoming a threat again in the future; there is now little temptation for Acrasia to return if the opportunity presented itself, and certainly Grill is in no position to rebuild the Bower.

Thus Guyon's actions suggest an effort at supplanting rather than at planting. Once he has accomplished his purpose, one which evidently accords with the Palmer's own understanding of Guyon's mission, there appears to be little left to do but depart. The same tactic was familiar to Drake, Hawkins, Cavendish, and Ralegh, all but the last of whom were more concerned with privateering than with colonization. The analogy is blurred, however, by Guyon's failure to behave like a pirate; he burns but does not pillage, leaving with nothing other than his two captives. In this account, stanza 83 displays Guyon neither as a corsair nor as a missionary, but as, so to speak, a bailiff. For all its lyrical power, the stanza may describe little more than a police action, the ramifications of which do not extend beyond the nine lines in which the action occurs.

Tactics are not identical to strategy, nor is police action identical to policy. It is difficult to accept stanza 83 as a limited solution to an immediate problem rather than as the fulfillment of a plan as large as Book 2 itself. But at the end of the book, this larger plan has yet to emerge in any clear form. Spenser finally offers no alternative to Guyon's essentially negative status in Book 2. While Guyon empties the Bower of its former meanings, he omits to fill it with meanings of his own. The pattern in the other books from 1590 involves conclusions where the eradication of an evil is followed by the establishment of a positive good signified by betrothal: Redcross and Una become engaged; Britomart restores Amoret to Scudamour. In contrast to Books 1 and 3, Book 2 ends with a strong sense of disengagement. The original motive for Guyon's quest—revenge for the deaths of Mordant and Amavia—has quite likely been forgotten by most of the poem's readers, who have no clear stake in the fates of Verdant and Acrasia and no clear idea of what the later judgment upon them will be. No renewed or "reformed" settlement rises from the ashes of the Bower; Spenser has reduced the text itself to "nought but desert wildernesse" (7.2.9).

Spenser's Temperance may seem a strangely vacant construct, where balance is achieved not by using counterweights but simply by dismantling the scales. Or, to return to the metaphor of tempered metal broached in chapter 4, it could be said that Guyon never receives a final hardening but remains a malleable figure in readers' minds. This is not to imply a flaw in Spenser's method; in fact the quality of uncertainty in Book 2 is entirely consistent with that method. Spenser has devised, in cumulative fashion, an elaborate critique of one manner of conducting empire. Critique is not the same as doctrine, however; whatever else he may be, Spenser is not a doctrinaire poet. Generations of his readers have struggled mightily to outline a consistent ideology within *The Faerie Queene*. This impossibility does not transform Spenser into a writer without ideology; a cursory glance at *A View of the Present State of Ireland* proves otherwise. It is worth recalling, though, that the *View* never circulated outside of manuscript during Spenser's lifetime, and that Spenser, rather than rendering a straightforward account of his opinions on Ireland, presented them by way of an elaborately artificial Platonic dialogue.

The sixteenth century in England, unlike much of the seventeenth century, was a period of veiled political discourse; the richness of Elizabethan literature at times conceals the fact that for Spenser and his peers, "free speech" was still the bedfellow of treason. No matter what Spenser thought, said, or did when he turned away from his muse, the great product of that muse could not belong to the world of action. It could not dispossess the Spaniards from the Indies or claim Virginia for England. Moreover, it could not tell Queen Elizabeth what to do. If Spenser chose to undertake such a task he would not have written an allegory; he would have written something more hazardous to his own career as a poet, civil servant, and beneficiary of royal patronage within what remained, for all its loopholes and forebearances, an absolute and often arbitrary monarchy. The limitations on the Elizabethan writer lend *The Faerie Queene* its "gestural" quality; it points to many areas of contemporary and national discourse—including a putative discourse of imperial activity—but cannot be said truly to belong within any one of those discourses, other than the traditional discourse of poetry. In addressing Book 2's proem at the beginning of this book, I talked metaphorically about Spenser holding up a looking glass that is also a window; however, *The Faerie Queene* never really offers a door though which its readers can pass to an English New World.

Sidney notoriously declares in *A Defence of Poetry* that "the poet, he nothing affirms, and therefore never lieth."[49] The implicit aspect of this argument is that the poet, for all his ability to please and instruct, is not precisely telling the truth. He is, Sidney says, "not labouring to tell you what is or is not, but what should or should not be."[50] The "shoulds" and "should nots" of *The Faerie Queene* are voiced conditionally rather than imperatively. And Spenser's critics, however politicized their practice may have become, will continue to find difficulty in either transforming or canceling out the persistently conditional "voice" of Spenser's allegory. Spenser states his own position by way of the Palmer; in a sense the Palmer's voice cannot be discriminated from Spenser's own: "But let us hence depart, whilest wether serues and wind."

Conclusion: Allegory and Agency

> For poetry makes nothing happen: it survives
> In the valley of its saying. . . .
>
> W. H. Auden, "In Memory of W. B. Yeats"

In the previous six chapters I have laid out a case for Spenser composing an allegory of English colonial conduct that draws upon a knowledge—a broader and more precise knowledge than has previously been attributed to him—of texts about the Spanish New World empire that were available in England from the 1550s onward. To an extent, I have been working through this as a problem of literary history: What were Spenser's sources and how did he employ them? What are the appropriate documentary and circumstantial contexts for Book 2? At the same time, this study—as with any study of Spenserian allegory—inevitably raises more rarefied questions about intent and effect and the web of relations between the author, the text, and the reader. Such questions persist in the instance of Book 2 because Spenser's approach to his narrative of temperance is itself so tempered; like Guyon, Spenser seems to refrain from the positive assertion, the conclusive gesture, or the arrival at a terminus or a synthesis. The conduct of the English colonist, in this reading, is all restraint, deference, reluctance—an avoidance of the model of empire provided in such a declarative form by the Spanish. The same strategy of restraint serves Spenser well, allowing him to postpone indefinitely a potential crisis, one which can be pictured this way: the conquistador and the anti-conquistador face each other in the heat of battle over new lands, and suddenly recognize an uncanny family resemblance.[1] The reader may feel that this moment of anagnorisis is imminent, but finds him- or herself still gazing into Spenser's "faire mirrhour," waiting (evidently without end) for an image of the future, an ultimate "Elizabethan projection," to emerge.

The problem raised by the apparently highly deliberate inconclusiveness of Spenser's design in Book 2 leads me back to the lingering question of his contribution to colonial discourse in the late Elizabethan years. If only a handful of people are willing to read his work, of course, then the question may well be moot. *The Faerie Queene* is in peril of joining Ralegh's *The History of the World* as a Renaissance masterwork that is not only unfinished but also largely unread. Unlike Shakespeare, Spenser has suffered considerably from the scrutiny he has received as a creature of Elizabethan politics and ideology; his fate has been to drop down the curricular hierarchy into the purgatorial region of "representative" Elizabethan literature inhabited by the Gascoignes and Lylys, the Daniels and Draytons—"food for powder," to borrow Falstaff's famous description of his humble foot soldiers; "they'll fill a pit as well as better."[2]

There is little point in trying to rescue Spenser's poetry from that pit for the benefit of present-day readers; such rescue operations happen (if they happen) in the classroom, and, in any case, most of Spenser's readers already know who they themselves are. But I will argue that important aspects of Elizabethan literature also go unread in the general move toward a historicized and politicized criticism of Renaissance culture (and it is hard to find any other kind of Renaissance criticism at the present time). It is possible to claim that culture is invariably political, that literature is no less political than anything else in a culture, and that in this sense authors are political creatures. It is possible, indeed easy, to view Spenser as complicit in the workings of a despotic colonial regime by virtue of his status as a landholder in Ireland, his authorship of *A View,* and his sometimes fulsome celebrations of Elizabeth's rule. Yet these possibilities fail to account for the distinctive character of *The Faerie Queene,* or for Spenser's status *as a poet* in the Elizabethan social order. Complicit as an English colonist of Irish land Spenser certainly was; but in his role as an author the case becomes much less obvious. I know of no evidence indicating that Spenser's writings, including *A View,* affected the course of Elizabethan affairs of state in any tangible way, or even that those writings influenced popular conceptions about Elizabeth or her policies.[3]

The argument can be made that *The Faerie Queene,* as an instance of the kind of "projection" outlined in chapter 1, contributes to a general reinforcement of certain cultural and political models promulgated by various social institutions (or, in Louis Althusser's now-familiar coinage, ideological state apparatuses[4]). But this argument is almost always made collectively (i.e., "*The Faerie Queene* along with many other works . . ."), with no work bearing greater responsibility than any other in the enterprise of reification. This would appear to be the approach that Richard Helgerson takes, for example, when he claims in the introduction to *Forms of Nationhood* that "Texts, nations, individual authors, particular discursive communities—all are both produced and productive, productive of that by which they are produced."[5] In this sense Spenser's work would be confluential rather than influential, part of the massive flow of information that constitutes both history and culture. And, as Spenser individuates the many rivers that attend upon the marriage of the Thames and Medway in Book 4 of the poem, so the literary historian breaks down the confluence into its tributary sources, both major and minor, as one way of gaining an understanding of the whole. There is little need for the literary historian to construct teleological chains out of these individual elements in order to explain the existence of the confluence itself, except at the highest level of generality: i.e., many things coming together make a bigger thing.[6]

Assuming that *The Faerie Queene* is an element in such a confluence, what kind of work does it do there, other than contribute to the total flow? In chapter 1, I suggested that the motivating force behind projection is the desire to make the text become the world, to make textual activities become worldly actions, while somehow leaving behind the task (or burden) of representation that is inseparable from writing itself. It is possible to see *The Faerie Queene* as participating in such an idealistic and implausible undertaking, but with the marked difference that the poem constantly advertises its allegorical status, therefore its attenuated and mystified relation to the things of this world. Spenser informs his readers at the very outset, by way of the letter to Ralegh, that the "good discipline" of the poem is "cloudily enwrapped in Allegorical devices." He also acknowledges that his poem cannot impose its vision of the world on the world as it is; he is constrained by the superficial quality of human understanding (which esteems of "nothing . . . that is not delightful and pleasing to common sense") to follow the model of Xenophon rather than of Plato, "for that . . . [Plato] in the exquisite depth of his judgment, formed a Common wealth such as it should be, but the other in the person of Cyrus and the Persians fashioned a government such as might best be: So much more profitable and gracious is doctrine by example, than by rule."[7] Plato's vision of the republic may be preferable from Spenser's perspective, but it is both too demanding and too deep for the world to embrace. Doctrine by example (that is, by representation) is more effective because it is more palatable to weak and erring humankind—except, of course, that examples need not be followed and, even if they are followed, the doctrines behind them need not be fully absorbed. The characteristic double bind in Spenser's allegory (perhaps in allegory generally) has to do with its highly inefficient didacticism: it can never dictate its teachings to the world, but at the same time it risks being misunderstood—or at least incompletely understood—by whispering to the world instead. Due to its primary poetic mode, *The Faerie Queene* cannot present or promote its "project" in the way that many Elizabethan political and mercantile tracts attempt to do.

Yet the reader is always conscious that a project of some sort is going on in Spenser's poem, that a picture of the world is being rendered, ideas are being put to the test, and connections are being drawn between things concrete and abstract. Allegorists undertake these activities no less than other writers. What distinguishes allegory from other modes of writing is its constant acknowledgment of the distances that must be covered in making these connections, in bringing the world into the text as well as the text into the world. It would even be plausible to say that allegory represents these very distances. A poem like *The Faerie Queene* demonstrates, indeed celebrates, the sheer difficulty of superimposing the ideal on the real, and exposes the quantity of illusion and equivocation involved in trying to do so. Spenser is in this sense like a magician who shows the technique behind the trick before performing his sleight of hand. The inherent paradox in allegory—that it can be deeply mysterious in its content while being openhanded about the rules of its craft—is what gives it its intellectual savor and its value as entertainment. Its emotional power, on the other hand, emerges at those moments, such as the expostulation on the "fourth Mayd" of Mount Acidale in Book 6 canto 10 (25–28), when the distance between the real and the ideal abruptly seems to be bridged and the world seems to become directly present to both author and audience. The extraordinary poignancy of such moments comes from the fact that they are, in the vast expanse of Spenser's poem, so hard-won and so transient.

One function of a work like *The Faerie Queene,* then, is to raise a difficulty—a difficulty that relates to conflict between the human need to express the world (as "all that is the case," to borrow Wittgenstein's famous definition[8]) and the fundamental inexpressibility of that world. This is not far off from the notion, familiar in new historicist criticism, that literary texts are primarily records of the large- and small-scale tensions in a given cultural milieu. I agree with Greenblatt's claim that "great art is an extraordinarily sensitive register of the complex struggles and harmonies of culture."[9] The problem arises when attempts are made to position this register within a realm of ethical and political consequences, as Brook Thomas points out in his incisive critique of Greenblatt's understanding of mimesis. The "metaphor of a register is useful," he says. "A device used to measure an earthquake can do so because it is somehow connected to the thing it records. Nonetheless, it would be very hard to measure the effect of the register on the earthquake."[10] The problem is compounded in the case of allegory, which almost always presents itself as an *indirect* register of social and cultural movements. Thus, allegory exaggerates what Thomas sees as the leading feature of literary texts in general; drawing on the work of Wolfgang Iser and Hans Blumenberg, Thomas argues that such texts are far more likely to deform than reform the world: "The literary text is not so much characterized by its re-presentation of a world in its fullness as by its negativity, which keeps the various discursive systems appropriated by the text from uniting into a balanced organic harmony." The content of the text is not "an extension of any one particular discursive system" but "is, as it were, a new language that resists translation into any of those that helped bring it into existence. This resistance to translation is the literary text's strength and weakness." The text's weakness arises "because, by resisting translation, it seems incapable of circulation."[11] Thomas's formulations strike me as coming close to the heart of the matter, especially in regard to a work like *The Faerie Queene.* Since Spenser literally offers his readers "a new language that resists translation," he forestalls the possibility that his poem will circulate as, say, Las Casas's *Brevissima Relación,* Martyr's *Decades,* or Hakluyt's *Principall Navigations* could circulate—that is, in any sense in which the political and the literary can be said to intersect significantly.

Yet all literary works that are not simply private documents must circulate in some fashion, however limited; *The Faerie Queene* has always had its "public," if never a very large one. This brings me to the question of the poem's agency, which is, finally, a question of how readers respond to the poem, for it is difficult to talk about agency in the absence of reaction. Here again one faces the problem of accurately assessing the contemporary (or retrospective) impact of Spenser's work on Queen Elizabeth's reputation, England's national pride, the habits and tastes of readers, political or religious sensibilities, and any collective activity undertaken by the English in either the Old World or the New. But Spenser's readers have nearly always understood his work as exemplifying an approach to poetry characterized by deep erudition, high (if also highly self-conscious) artifice, and meticulous attention to form. *The Faerie Queene*'s influence, as such, lies with the transmission of that approach to other texts by other writers. It is within the confines of literary history, so to speak, that the poem circulates most effectively. And how many poems ever written can be understood as having an impact above and beyond that history?

This is not to deny the power of Spenser's poetry or of poetry generally, but simply to state the limitations of such writing to range freely among other kinds of discourse in a culture. Given the constraints on its circulation, *The Faerie Queene* nonetheless continues to function as a register, or perhaps more appropriately a testimony, to a difficulty.[12] Thomas sees this difficulty manifesting itself as paradox: "Literary texts generate paradoxes, not because they successfully represent a world of plentitude above history, but because they are constituted by a lack, a space of negativity allowing readers to participate in their discursive play."[13] The reader's consciousness of this lack corresponds to what Gerald L. Bruns, in defense of Hans-Georg Gadamer's theory of hermeneutics, describes as the experience of "openness," which "is a condition of exposure in which one's conceptual resources have been blown away by what one has encountered."[14] Bruns sees such a "condition" as being endemic to acts of textual interpretation:

> In our encounter with a text that comes down to us from the past (or from another culture, or perhaps from our very midst), we always find that what the text says is not what we thought (or what we would say) but something else, something other that brings us up hard against the presuppositions or forestructures that we inhabit and that underwrite and make possible our efforts of understanding.[15]

What readers come to recognize when they participate in the paradoxical movements of literature—when they find themselves, so to speak, in an exposed position—is what Thomas, again paraphrasing Iser, calls "the embarrassing predicament of the failure of our understanding."[16] A further paradox is that what is embarrassing is usually also stimulating. I would claim that in *The Faerie Queene* Spenser offers an intricately designed, exquisitely modulated, historically situated account of this very predicament.

Throughout this book I have explored the problems inherent in trying to define a model of empire by way of negation. To generalize further, I would say that literary activity typically involves attempting to define all manner of worldly things in relation to "a lack," or "a space of negativity." I have already suggested that such attempts produce a consciousness of distance which, in a work like *The Faerie Queene*, cannot be reduced to disinterest, ironic detachment, or a vantage point on some moral, ethical, or spiritual high ground. This distance is, simply but centrally, the difficulty that the poem raises, and it is one that readers, writers, critics, and scholars can inherit and preserve within a "literary tradition," if it is still possible to speak of such a thing. For it is mainly within such a tradition that questions can be asked with a simultaneous awareness that they are largely unanswerable. In *The Faerie Queene* Spenser asks many unanswerable questions: Where is Fairy Land? What sort of empire does Gloriana rule? What is Temperance? What is a "gentleman or noble person"?[17] Paradoxically, he also presents his audience with a well-defined, magisterially constructed means for doing the asking. It is here, and not in the poet's actual or alleged influence over Elizabethan policy, that Spenserian agency finally makes itself known.

NOTES

Introduction

1. See the editors' introduction to *A View of the State of Ireland,* ed. Andrew Hadfield and Willy Maley (Oxford: Blackwell, 1997) xi–xii, xxiv–xxvi.
2. Willy Maley, *Salvaging Spenser: Colonialism, Culture and Identity* (New York: St. Martin's, 1997), 90.
3. Richard A. McCabe, "Edmund Spenser, Poet of Exile," *Proceedings of the British Academy* 80 (1993): 74.
4. Stephen Greenblatt, *Renaissance Self-Fashioning: From More to Shakespeare* (Chicago: U of Chicago P, 1980), 157–92.
5. But it is worth noting that it was Edwin Greenlaw, not Stephen Greenblatt, who first placed "Spenser" and "British Imperialism" in close proximity, more than eighty years ago. See Edwin Greenlaw, "Spenser and British Imperialism," *Modern Philology* 9 (1912): 347–70. See also Lois Whitney, "Spenser's Use of the Literature of Travel in the *Faerie Queene,*" *Modern Philology* 19 (1921): 143–62; Robert Ralston Cawley, *Unpathed Waters: Studies in the Influence of the Voyagers on Elizabethan Literature* (Princeton: Princeton UP, 1940); and Roy Harvey Pearce, "Primitivistic Ideas in the *Faerie Queene,*" *Journal of English and Germanic Philology* 44 (1945): 139–51. More recent studies which predate the New Historicism include A. Bartlett Giamatti, "Primitivism and the Process of Civility in Spenser's *Faerie Queene,*" Fredi Chiappelli, ed., *First Images of America: The Impact of the New World on the Old,* 2 vols. (Berkeley: U of California P, 1976) 1:71–82; Thomas H. Cain, *Praise in "The Faerie Queene"* (Lincoln: U of Nebraska P, 1978); and Barbara-Maria Bernhart, "Imperialistic Myth and Iconography in Books I and II of the Faerie Queene," Ph.D. diss., McMaster University, 1975.

6. Not all of the works on this list would claim a direct debt to Greenblatt, but their authors would probably acknowledge that the acceptance of Greenblatt's work gave them the latitude to work on their own related projects. See Peter Hulme, *Colonial Encounters: Europe and the Native Caribbean, 1492–1797* (London: Methuen, 1986); Mary B. Campbell, *The Witness and the Other World: Exotic European Travel Writing 400–1600* (Ithaca, NY: Cornell UP, 1988); Eric Cheyfitz, *The Poetics of Imperialism: Translation and Colonization from The "Tempest" to "Tarzan"* (New York: Oxford UP, 1991); Richard Helgerson, *Forms of Nationhood: The Elizabethan Writing of England* (Chicago: U of Chicago P, 1992); Jeffrey Knapp, *An Empire Nowhere: England, America, and Literature from "Utopia" to "The Tempest,"* The New Historicism: Studies in Cultural Poetics 16 (Berkeley: U of California P, 1992); John Gillies, *Shakespeare and the Geography of Difference,* Cambridge Studies in Renaissance Literature and Culture 4 (Cambridge: Cambridge UP, 1994); Frank Lestringant, *Mapping the Renaissance World: The Geographical Imagination of the Age of Discovery,* trans. David Fausett, intro. Stephen Greenblatt, The New Historicism: Studies in Cultural Poetics 32 (Berkeley: U of California P, 1994) and *Cannibals: The Discovery and Representation of the Cannibal from Columbus to Jules Verne,* trans. Rosemary Morris, The New Historicism: Studies in Cultural Poetics 37 (Berkeley: U of California P, 1997); William M. Hamlin, *The Image of America in Montaigne, Spenser, and Shakespeare: Renaissance Ethnography and Literary Reflection* (New York: St. Martin's, 1995); Mary C. Fuller, *Voyages in Print: English Travel to America, 1576–1624,* Cambridge Studies in Renaissance Literature and Culture 7 (Cambridge: Cambridge UP, 1995); J. Martin Evans, *Milton's Imperial Epic: Paradise Lost and the Discourse of Colonialism* (Ithaca, NY: Cornell UP, 1996); Gesa Mackenthun, *Metaphors of Dispossession: American Beginnings and the Translation of Empire, 1492–1637* (Norman: U of Oklahoma P, 1997); Walter S. H. Lim, *The Arts of Empire: The Poetics of Colonialism from Ralegh to Milton* (Newark: U of Delaware P, 1998); Joan Pong Linton, *The Romance of the New World: Gender and the Literary Formations of English Colonialism,* Cambridge Studies in Renaissance and Culture 27 (Cambridge: Cambridge UP, 1998); and Shannon Miller, *Invested with Meaning: The Raleigh Circle in the New World* (Philadelphia: U of Pennsylvania P, 1998). Greenblatt's more recent meditations on colonialism can be found in *Marvellous Possessions: The Wonder of the New World* (Chicago: U of Chicago P, 1991).

   I would be remiss not to mention the now largely forgotten ur-text in this genre, one which predates Greenblatt's writings by quite a few years: Howard Mumford Jones, *O Strange New World: American Culture: The Formative Years* (New York: Viking, 1964).
7. Greenblatt, *Renaissance Self-Fashioning,* 174.
8. It is not that critics have abandoned these waters completely, but those who continue to trawl there all seem concerned to a degree with keeping out of Greenblatt's wake. Jeffrey Knapp, for example, pursues more emphatically than Greenblatt the notion that Spenser is contrasting "imperial" England with imperial Spain: "If any of her subjects helped Elizabeth to represent herself as a conqueror more benign and therefore more powerful than the king of Spain," he argues, "it was Spenser" (Knapp, *Empire Nowhere,* 15). Knapp also

notes Spenser's characteristic "process of self-definition by negatives" (128). But Knapp appears intent on avoiding another visit to the Bower of Bliss, opting instead for a reading of the Error episode in Book 1 canto 1 as an allegory of the inadequacies of Elizabethan imperialism, a reading involving an enormous degree of extrapolation from a passage that offers little direct—or even indirect—reference to colonial activity (106–33). In attempting to distance his argument from Greenblatt's, Knapp also manages to defer to it: Spenser is still a poet of empire under the terms established in *Renaissance Self-Fashioning* (see Knapp, *Empire Nowhere,* 287n. 1).

Another strategy is represented in Shannon Miller, 26–49, who *does* focus upon the proem and canto 12 of Book 2 as potential windows on Spenser's colonialism, but argues the basic absence of "real" New World references in the poetic design: "Spenser seems . . . to create a New World voyage that is not a New World voyage" (46). The underlying itinerary of Guyon's journey in this account is away from idle maritime projects (allegorized in Phaedria's wayward ferrying around the Idle Lake) toward a more practical as well as accessible colonial goal—Ireland. Thus Miller responds to Greenblatt's interpretation by, in effect, reversing direction and joining the scholarly fleet now fishing in Irish waters. My argument in this book, of course, begins with the notion that Greenblatt was headed toward the right compass point but that (perhaps because *The Faerie Queene* was only of passing interest to him) he left much of the most interesting and important territory around the Bower virtually unexplored.

9. See Greenblatt, "Learning to Curse: Aspects of Linguistic Colonialism in the Sixteenth Century," Chiappelli, *First Images of America,* 2:561–80. This essay is collected as well in Greenblatt, *Learning to Curse: Essays in Early Modern Culture* (New York: Routledge, 1990), 16–39. See also Greenblatt, *Shakespearean Negotiations: The Circulation of Social Energy in Renaissance England* (Berkeley: U of California P, 1988), 129–63.
10. "Greenblatt's Spenser," as Jay Farness has trenchantly observed, "is a personification—a type, a concept of what Guyon and Spenser are supposed to share, a concept that is reconstructed from selected circumstances of the text and, incidentally, represented with such energy, conviction, and likelihood that Greenblatt's Spenser has probably become more popular than its primary text" (Anderson et al., *Spenser's Life,* 22).
11. See, for example, the passing remarks on the "bloodthirsty" essence of *A View* made in Edward Said's *Culture and Imperialism* (New York: Knopf, 1993), 7, 222, 236.
12. Andrew Hadfield, *Edmund Spenser's Irish Experience: Wilde Fruit and Salvage Soyl* (Oxford: Clarendon, 1997), 12.
13. In addition to Hadfield, Maley, and McCabe, see David J. Baker, *Between Nations: Shakespeare, Spenser, Marvell, and the Question of Britain* (Stanford: Stanford UP, 1997); Jean R. Brink, "Constructing the *View of the Present State of Ireland,*" *Spenser Studies: A Renaissance Poetry Annual* 11 (1990): 203–28; Christopher Highley, *Shakespeare, Spenser, and the Crisis in Ireland,* Cambridge Studies in Renaissance Literature and Culture 23 (Cambridge: Cambridge UP, 1997); Lim, *Arts of Empire,* 142–93; S. Miller, *Invested with Meaning,* 50–

85; and a number of the essays in *Representing Ireland: Literature and the origins of conflict, 1534–1660,* ed. Brendan Bradshaw, Andrew Hadfield, and Willy Maley (Cambridge: Cambridge UP, 1993), esp. Hadfield and Maley's introduction, "Irish Representations and English Alternatives" (1–23); Lisa Jardine, "Encountering Ireland: Gabriel Harvey, Edmund Spenser, and English Colonial Ventures" (60–75); and Julia Reinhard Lupton, "Mapping Mutability: or, Spenser's Irish Plot" (93–115).

14. There are few satisfactory accounts of the Smerwick massacre, and the available accounts frequently fail to agree on basic facts about the event. The discussion in Alexander C. Judson, *The Life of Edmund Spenser* (Baltimore: Johns Hopkins UP, 1945), 89–93 is still adequate, though Judson nowhere mentions Ralegh's involvement. Among the most detailed accounts is Richard Berleth, *The Twilight Lords: An Irish Chronicle* (New York: Knopf, 1978), 161–76, but even here there are nagging problems regarding the historical record. Berleth says that the Fort del Oro received its name because one of the returning ships from Frobisher's 1577 voyage to Meta Incognita (i.e., Newfoundland) had foundered on the headlands there and dumped its cargo of fool's gold from the "mines" that Frobisher had discovered in the New World. This story dovetails nicely with a discussion of Spenser's colonial interests; however, I have found no evidence in the contemporary material on Frobisher's voyages that this shipwreck (assuming that it occurred) involved one of Frobisher's ships. The three ships on the second voyage in 1577 all returned safely to port. Of the fifteen sail on the third voyage in 1578, one was lost to pack ice off of Newfoundland and another returned home early without authorization (there is no mention of this ship's having wrecked on the homeward journey). The remaining ships all eventually made port in England. See Richard Collinson, ed., *The Three Voyages of Martin Frobisher,* Hakluyt Soc. 38 (London, 1867), 117–363. Berleth seems to have gotten this story from at least one biography of Ralegh; it appears in Norman Lloyd Williams, *Sir Walter Ralegh* (London: Eyre and Spottiswoode, 1962), 29, but without any attribution. Under the circumstances the story has to be regarded as apocryphal. All of this is to say that the episode of the Smerwick massacre deserves more meticulous archival research than it has received.
15. Spenser, *A View,* 104–5.
16. Ibid., 104.
17. S. Miller, in *Invested with Meaning* (50–85), makes the case for the special bearing of Ireland upon Spenser's understanding of Anglo-Spanish relations: "Ireland is not simply another colony in a potential dispute: it is *the* colony that validates England's ability to compete with Spain throughout known and unknown worlds" (70). But because there is so much emphasis in her discussion on Ireland as a sort of filter through which almost any English colonial project must pass, the possibility that Spenser might have a more "global" interest in Spanish colonial activity receives relatively little attention.
18. Leicester and Sidney show up in the lists of venturers for all three of Frobisher's voyages, and Dyer does for the second and third. See Collinson, *Martin Frobisher,* 107–9, 224, 346, 348–49, 351–52 (where John Dee himself appears as a contributor). For a thorough recent discussion of Spenser's connections

with Leicester and the Sidneys and the impact of these connections on his employment with Lord Grey, see Vincent P. Carey and Clare R. Carroll, "Factions and Fictions: Spenser's Reflections of and on Elizabethan Politics," Judith H. Anderson et al., eds., *Spenser's Life and the Subject of Biography* (Amherst: U of Massachusetts P, 1996), 31–44. In the same volume, see Jean R. Brink, " 'All his minde on honour fixed': The Preferment of Edmund Spenser" (45–64) for an analysis of Gabriel Harvey's possible status as an intermediary in this same network of relationships.

19. Charles Nicholl, *The Creature in the Map: A Journey to El Dorado* (New York: Morrow, 1995), 38.
20. Gordon Teskey, *Allegory and Violence* (Ithaca, NY: Cornell UP, 1996), 170.
21. Teskey, *Allegory and Violence,* 174.
22. A. C. Hamilton, ed., *Faerie Queene,* 737. I have modernized Spenser's spelling here and in subsequent quotations from the letter to Ralegh.
23. Michael O'Connell, "allegory, historical," *The Spenser Encyclopedia,* ed. A. C. Hamilton (Toronto: U of Toronto P, 1990).
24. Baker, *Between Nations,* 8.
25. Keith Robbins, *Great Britain: Identities, Institutions and the Idea of Britishness* (London: Longman, 1998), 120.
26. The best recent account of the complexities of the early British "empire" is Kenneth R. Andrews, *Trade, Plunder and Settlement: Maritime Enterprise and Genesis of the British Empire, 1480–1630* (Cambridge: Cambridge UP, 1984).
27. René Graziani, "*The Faerie Queene,* Book II," *The Spenser Encyclopedia,* 263.
28. Thomas H. Cain, "New World," *The Spenser Encyclopedia,* 510.
29. This is the point of departure for Knapp's *An Empire Nowhere,* which may well represent the last word in research on the literary-historical context of Elizabethan colonialism; Knapp seems literally to have examined every available source in this area. But this exhaustive approach is tied to an extravagantly associative and circumstantial style of argumentation which—appropriately, given the book's emphasis on the significance of Virginian tobacco—often allows the smoke to exceed the fire. I am aiming here for a much more circumscribed but also, I hope, more measured approach to some of the same problems that Knapp considers.
30. I agree in principle with Jay Farness's argument for a form of Spenserian interpretation that "in the general rush to history, hesitates a little longer at the possibility of something more specifically literary, ways of reading that have not wholly abandoned New Critical or new Critical practices, in part because they are not yet sure just what Spenser's literary text is or what it is capable of doing. For such readers *The Faerie Queene* continues to present a literary opacity that does not readily translate into historical or biographical information, readers for whom old- and new-historical studies are deeply interesting but inadequate as 'Spenser interpretation.' This more literary focus on the poetic processing of theme tends to see a poet not submerged by discursive resources but continuing to operate in, with, and through them" (Anderson et al., *Spenser's Life,* 28).
31. I am thinking especially here of Cain's treatment of Book 2 in *Praise in "The Faerie Queene"*: While Cain predates Greenblatt in elaborating the links between Book 2 and Elizabethan imaginations of American empire, and while

he is quite sensitive to the influence of the Spanish model on English thinking, he finally transforms Book 2 into a parable of Ralegh's search for El Dorado, which he bases in part on a perceived etymological relation between "Guiana" and "Guyon" (96–97). The difficulty with this reading, of course, is that Ralegh's voyage was still five years in the future when Spenser published the first three books of *The Faerie Queene,* so that the question is not how Ralegh influenced Spenser, but how Spenser influenced Ralegh. (On the subject of Spenser's influence on Ralegh in Guiana, see the informed speculations in Nicholl, *Creature in the Map,* 304–13.) See also Cain's article, "New World," *The Spenser Encyclopedia,* 510. The dissertation of Cain's student Bernhart exhibits a similar historical reductionism at much greater length.

32. A. C. Hamilton, *Faerie Queene,* 737.
33. Helgerson, *Forms of Nationhood,* 54.
34. Helgerson ultimately says as much, in one of the last footnotes in *Forms of Nationhood:* "The Faerie Queene not only upsets the categories that control most of the other major textual representations of England to emerge from Spenser's generation, it also upsets its own categories" (350).

Chapter 1

1. The essential optimism of Spenser's approach here dissipates in the proems to Books 5 and 6, where antiquity carries the day over the decadent present and its prospects. But each proem must be considered in the context of the book that it introduces. I would also note that in every instance the virtues of the Queen exceed those of her forebears in antiquity.
2. For another discussion of Spenser's prophetic stance in the proem see Cain, *Praise,* 85. See also S. Miller, *Invested with Meaning,* 34–37, 81, who makes the puzzling claim that the proem is "the least frequently discussed reference to New World discoveries in the poem" (34).
3. Greenblatt, in *Renaissance Self-Fashioning* (190–91), proposes that this "faire mirrhour" only offers a reflection of Elizabeth herself: "the queen turns her gaze upon a shining sphere hitherto hidden from view and sees her own face, her own realms, her own ancestry" (191). My argument in the chapters that follow is that the reflections in Spenser's mirror are anything but definitive.
4. Or at the very least that of an apostle of the new, willing to apply the innovations of Giordano Bruno to the problem before him: "What if in euery other starre vnseene / Of other worldes he happily should heare? / He wonder would much more: yet such to some appeare" (Pr.3.7–9). For Bruno's ideas on the infinitude of worlds see Frances A. Yates, *Giordano Bruno and the Hermetic Tradition* (1964; reprint, Chicago: U of Chicago P, 1979), 235–56.
5. *The Letters of Amerigo Vespucci and Other Documents Illustrative of His Career,* trans. Clements R. Markham, Hakluyt Soc. 90 (London, 1894). See esp. 42: "and it is lawful to call it a new world, because none of these countries were known to our ancestors, and to all who hear about them they will be entirely new." The irony here is that the land that Vespucci claimed to have discovered in 1501 (Brazil seems to be the point of reference, though the confusing and contradictory navigational information in the letter to Lorenzo di Medici makes it difficult to know for certain) had already been discovered in 1499, during the respective voyages of Vicente Yáñez Pinzón and the Portuguese Pedro Alvares

Cabral. Thus it was not "entirely new." Yet Vespucci's language in the Medici letter—quite likely presenting the first instances in which these lands were denominated as "new," and not as appendages of the Orient—ultimately had a profound effect on the way Europeans conceived of the Americas. Markham's introduction to the *Letters* (i–xliv) briefly and effectively addresses the problems surrounding Vespucci's career and literary legacy.

6. Bartolomé de las Casas, *History of the Indies,* trans. and ed. Andrée Collard (New York: Harper, 1971), 237.
7. The *Oxford English Dictionary* offers as its first example of this meaning a citation from John Dee's 1570 preface to Billingsley's Euclid.
8. Fulke Greville, *Life of Sir Philip Sidney* (Oxford: Clarendon, 1907), 77.
9. For analyses of a number of these texts and their English translations, see William S. Maltby, *The Black Legend in England: The Development of Anti-Spanish Sentiment, 1558–1660* (Durham, NC: Duke UP, 1971), 12–28. For interesting recent discussions of Hakluyt see Helgerson, *Forms of Nationhood,* 151–91, and Fuller, *Voyages in Print,* 141–74; see also Maltby, *Black Legend in England,* 61–75.
10. The letters of Christopher Columbus and of Hernan Cortés, for instance, were curiously never translated into English, though they went through many editions on the continent.
11. This list is drawn from E. G. R. Taylor, *Tudor Geography 1485–1583* (London: Methuen, 1930), 193–243. Taylor's is an essential study. For a general discussion of the library see Peter French, *John Dee: The World of an Elizabethan Magus* (1972; London: Ark-Routledge, 1987), 40–61. A brief but classic evaluation of Dee's importance appears in Frances A. Yates, *Theatre of the World* (Chicago: U of Chicago P, 1969), 1–19. More recent studies of Dee include Nicholas Clulee, *John Dee's Natural Philosophy: Between Science and Religion* (London: Routledge, 1988) and William H. Sherman, *John Dee: The Politics of Reading and Writing in the English Renaissance* (Amherst: U of Massachusetts P, 1997).
12. Though there are various short documents written by Drake himself, the main accounts of his exploits are all the work of other hands such as Philip Nichols and Francis Fletcher. The most famous of these accounts, *Sir Francis Drake Revived* and *The World Encompassed by Sir Francis Drake,* do not appear in print until the late 1620s, though at least part of the former text was extant as early as 1592. Both are sponsored—perhaps also compiled—by Drake's nephew and namesake Sir Francis Drake. A convenient gathering of relevant Drakeana is found in John Hampden, ed., *Francis Drake Privateer: Contemporary Narratives and Documents* (London: Eyre Methuen, 1972).
13. The best edition of this work continues to be that of V. T. Harlow (London: Argonaut, 1928). Harlow's introduction, xv–cvi, provides an incisive critical account of Ralegh's complicated life history, particularly as it is revealed in his "discoveries."
14. Greenblatt's *Sir Walter Ralegh: The Renaissance Man and His Roles* (New Haven: Yale UP, 1973) argues that Ralegh was able to view his own career and reputation in a self-consciously theatrical way, as something that could be deliberately shaped to create aesthetic impressions in an "audience."
15. Fulke Greville, *Life of Sir Philip Sidney* (Oxford: Clarendon, 1907), 70–78.

16. Ibid., 109.
17. Ibid., 119.
18. Cited in Taylor, *Tudor Geography,* 114.
19. See Francis A. Yates, *Astraea: The Imperial Theme in the Sixteenth Century* (1975; reprint, London: Ark-Routledge, 1985), esp. 1–87. For a succinct analysis of the crucial phrase see G. R. Elton, *England Under the Tudors,* 2d ed. (London: Methuen, 1974).
20. E. G. R. Taylor, ed., *The Original Writings & Correspondence of the Two Richard Hakluyts,* 2 vols., Hakluyt Soc. 2d. ser. 76–77 (London: Hakluyt Soc., 1935), 2:332. Helgerson, in *Forms of Nationhood* (167), comments on a more expansive version of Hakluyt's list from the same document.
21. Taylor, *Original Writings,* 2:334.
22. Ibid., 360. Taylor's translation is as follows: "there yet remain for you new lands, ample realms, unknown peoples; they wait yet, I say, to be discovered and subdued, quickly and easily, under the happy auspices of your arms and enterprise, and the sceptre of our most serene Elizabeth, Empress—as even the Spaniard himself admits—of the Ocean" (367).
23. O. B. Hardison, Jr., ed., *English Literary Criticism: The Renaissance* (New York: Goldentree-Appleton, 1963), 171–72.
24. Edward Arber, ed., *The First Three English Books on America [1511]–1555 A.D.* (London, 1895), 50.
25. Arber, *English Books,* 55. The comparison of the English to "sheepe" is telling, since the "one trade" that dominated the English economy at the time was the manufacture of woolens.
26. See the introduction in Arber, xxxix–lx. For a fuller account of Eden's life and achievements, see John Parker, *Books to Build an Empire: A Bibliographical History of English Overseas Interests to 1620* (Amsterdam: N. Israel, 1965), 36–53. A recent consideration of the complexities of Eden's editorial role appears in Andrew Hadfield, "Peter Martyr, Richard Eden and the New World: Reading, Experience, and Translation," *Connotations* 5 (1995–96): 12–18.
27. The joint-stock Muscovy Company was established in 1555, and its agents, most notably Anthony Jenkinson, initiated successful trade with both Russia and Persia during the 1560s. As this trade waned due to constant turmoil in central Asia, the Levant Company, founded in 1581, picked up the slack through its very profitable spice-for-woolens trade in Turkey and points farther east. For a short history of these operations see Boies Penrose, *Travel and Discovery in the Renaissance, 1420–1620* (1952; reprint, New York: Atheneum, 1962), 239–58.
28. Thomas Nicholas, *The Pleasant historie of the Conquest of the VVeast India, now called newe Spayne, Atchieued by the vvorthy Prince Hernando Cortes marques of the valley of Huaxacac, most delectable to Reade: Translated out of the Spanishe tongue, by T. N. Anno. 1578. Imprinted at London by Henry Bynneman,* sig. a2r–v.
29. Ibid., sig. a4r.
30. Drake was in the midst of his circumnavigation in 1578, and thus incognito. Most likely Ralegh is intended; Sidney is also a possibility. The mention of "valiant beginnings" seems to suggest that the gentleman is still young.

31. Nicholas, *Pleasant Historie,* sig. b2r.
32. Nicholas, *The strange and delectable History of the discouerie and Conquest of the Prouinces of Peru, in the South Sea. And of the notable things which there are found: and also of the bloudie ciuill vvarres vvhich there happened for gouernment. Written in foure bookes, by Augustine Sarate, Auditor for the Emperour his Maiestie in the same prouinces and firme land. And also of the ritche Mines of Potosi. Translated out of the Spanish tongue, by T. Nicholas. Imprinted at London by Richard Ihones, dwelling ouer against the Fawlcon, by Holburne bridge. 1581.* sig. A4v.
33. Richard Hakluyt, *The Principall Navigations, Voiages and Discoveries of the English Nation,* 2 vols., Hakluyt Soc. Facsimile Ed. (Cambridge: Cambridge UP, 1965) 2:679.
34. Hakluyt, *Principall Navigations,* 2:680.
35. For another assessment of Hayes's account of Gilbert's voyage, see Fuller, *Voyages in Print,* 33–38.
36. Helgerson, *Forms of Nationhood,* 182–83. Helgerson argues that Hakluyt's solution to this problem is to present the figure of the English merchant as the antithesis to the conquistador. The argument could not be made in the same way about *The Faerie Queene,* since merchant figures apparently do not signify in Fairyland (181–87).

Chapter 2

1. A. C. Hamilton, *Faerie Queene,* 738.
2. This is not the case in the other books from 1590; the provenances of Redcross's and Britomart's respective quests from *within* Gloriana's court are clearly stated very early on (1.3 in Book 1 and 1.8 in Book 3).
3. Arber, *English Books,* 189.
4. See Charles Gibson, ed., *The Black Legend: Anti-Spanish Attitudes in the Old World and the New* (New York: Knopf, 1971), 3–27 for a short introduction to the growth of *la leyenda negra* and the modern response to it. For a detailed account specific to England, see Maltby, *Black Legend in England.*
5. *The Spanish Colonie, or Briefe Chronicle of the Acts and gests of the Spaniardes in the West Indies, called the newe World, for the space of xl. yeares: written in the Castilian by the reverend Bishop Bartholomew de las Casas or Casaus, a Friar of the order of S. Dominicke. And nowe first translated into english, by M. M. S. Imprinted at London [by Thomas Dawson] for William Bro[o]me. 1583,* sig. G1v. The *Dictionary of National Biography* tentatively identifies the translator as one Mark Sadlington, a divine with connections to Walsingham. This identification is made less doubtful given the odd paucity of Christian names beginning with "M" during the Elizabethan period. For the relevant years, the *Dictionary of National Biography* lists only five names with first and last initials "M" and "S"—one of which belongs to Mary Sidney, countess of Pembroke.
6. In this passage Cain finds "a clear allusion to Dee's claim that Elizabeth's ancient progenitor Arthur held dominion in America" (Cain, *Praise,* 99). But see also Michael Murrin, *The Allegorical Epic: Essays in Its Rise and Decline* (Chicago: U of Chicago P, 1980), 137–39.
7. So immediate is the book, indeed, that it ends "abruptly" before its reader's own impending birth (10.68.2).

8. For yet another variation on these themes, see Guyon's tribute to the Fairy Queen at the beginning of canto 9: "My liefe, my liege, my Soueraigne, my deare, / Whose glory shineth as the morning starre, / And with her light the earth enlumines cleare; / Far reach her mercies, and her prayses farre, / As well in state of peace, as puissaunce in warre" (9.4.5–9). Arthur then tells Guyon that he has compassed the globe looking for this far-reaching queen: "Now hath the Sunne with his lamp-burning light, / Walkt round about the world, and I no lesse" (9.7.5–6).
9. Thomas Har[r]iot, *A Briefe and True Report of the New Found Land of Virginia: The Complete 1590 Theodore de Bry Edition,* intro. Paul Hulton (New York: Dover, 1972), 75; the Pict illustrations and captions follow on 76–85. For the complex provenance of this group of illustrations see Hulton's introduction, xii–xiii. Stephen Orgel attempts to place de Bry's Picts in the background of *The Tempest* in "Shakespeare and the Cannibals," *Cannibals, Witches, and Divorce: Estranging the Renaissance,* ed. Marjorie Garber, Selected Papers from the English Inst. (1985) n.s. 11 (Baltimore: Johns Hopkins UP, 1987), 43–44. S. Miller, in *Invested with Meaning* (56–69), discusses the group at some length as intended to prompt associations with Irish natives. This strikes me as fairly tenuous, since the Elizabethans inevitably understand the Picts as inhabiting the north of England and are much more likely to link them historically with Scotland—another troublesome regional *locus* of the "primitive"—than with Ireland.
10. Spenser elaborates on this point in *A View of the Present State of Ireland:* "another nation coming out of Spain, arrived in the west part of Ireland, and finding it waste, or weakly inhabited, possessed it: who whether they were native Spaniards, or Gauls, or Africans, or Goths, or some other of those Northern Nations which did over-spread all Christendom, it is impossible to affirm. . . . but that out of Spain certainly they came, that do all the Irish Chronicles agree" (45–46).
11. S. Miller, in *Invested with Meaning* (70–73), argues that Spenser in *A View* wants to deny that these emigrants from Spain were authentically Spanish; instead, they were Gauls or another ethnic group more closely associated with the English themselves. Thus Spenser can assert the priority of English over Spanish territorial claims in Ireland. I would say rather that Spenser understands the utility of an ambiguous identification, since such an identification allows the prospect of an Irish-Spanish alliance to appear to convey a genuine threat to English interests *and* to rest on factually flimsy premises demanding some form of intervention or correction.

Chapter 3

1. C. S. Lewis, *English Literature in the Sixteenth Century Excluding Drama,* Oxford History of English Literature 3 (Oxford: Oxford UP, 1954), 380.
2. For a game, if overdrawn, recent effort to sort out the geographical relations of Fairyland and Britain in terms of generic affiliations to romance and epic, see Wayne Erickson, *Mapping "The Faerie Queene": Quest Structures and the World of the Poem* (New York: Garland, 1996).
3. James Nohrnberg, *The Analogy of "The Faerie Queene"* (Princeton: Princeton UP, 1976), 352.

4. Murrin, *Allegorical Epic,* 137, 139. See also Murrin's article on "fairyland," *The Spenser Encyclopedia,* 296–98.
5. See Anne Treneer, *The Sea in English Literature from Beowulf to Donne* (London: UP of Liverpool, 1926); B. Nellish, "The Allegory of Guyon's Voyage: An Interpretation," *English Literary History* 30 (1963): 89–106; Kathleen Williams, "Spenser: Some Uses of the Sea and the Storm-Tossed Ship," *Research Opportunities in Renaissance Drama,* Report of the MLA Seminar 13–14 (1970–71): 139–40; Jerome S. Dees's entry on "ship imagery" in *The Spenser Encyclopedia,* 655–56; and S. Miller, *Invested with Meaning,* 37, 42. See also A. D. S. Fowler, "Emblems of Temperance in *The Faerie Queene,* Book II," *Review of English Studies* n.s. 11 (1961): 143–49.
6. Lewis, *Allegory of Love,* 338.
7. For one possible "mapping" of Marinell's and Florimell's adventures, see Daniel M. Murtaugh, "The Garden and the Sea: The Topography of *The Faerie Queene,* III," *English Literary History* 40 (1973): 325–38. See also Isabel E. Rathborne, "The Political Allegory of the Florimell-Marinell Story," *English Literary History* 12 (1945): 279–89, for an interpretation of Florimell and Marinell as "connected, in Spenser's political allegory, with the two aspects of England's imperial mission which bulked largest in his thought, namely the civilization of Ireland and the conquest of the Ocean Sea and the New World" (288). Rathborne makes an interesting, if questionable, identification of Marinell's human father, Dumarin, with Christopher Columbus.
8. When Guyon and the Palmer pass by Phaedria at 12.14–17, Spenser does seem to distinguish between her current location and her previous one: "That was the wanton *Phaedria,* which late / Did ferry him [i.e., Guyon] ouer the *Idle lake*" (12.17.1–2).
9. Harry Berger, Jr., *The Allegorical Temper: Vision and Reality in Book II of Spenser's "Faerie Queene,"* Yale Studies in English 137 (New Haven: Yale UP, 1957), 5. Berger locates the "sharp break" at the point of Guyon's faint at the end of canto 7.
10. A multiplicity noted in another context in Judith H. Anderson, "The Knight and the Palmer in *The Faerie Queene,* Book II," *Modern Language Quarterly* 31 (1970): 160–61. Anderson is also concerned with Guyon's gradual separation from the landscape of the early cantos, though her treatment of the problem is very different from mine.
11. Arthur's need for succor is also thematically important in terms of Guyon's relationship with the Palmer; I will return to this point in chapter 5.
12. Gibson, *Black Legend,* 47.
13. Taylor, *Original Writings,* 2:257. The actual title of the treatise is *A particuler discourse concerninge the greate necessitie and manifolde comodyties that are like to growe to this Realme of Englande by the Westerne discoveries lately attempted.*
14. *The Works of Thomas Nashe Edited from the Original Texts,* ed. Ronald B. McKerrow (London: Sidgwick & Jackson, 1910) 1:178.
15. Mary Frear Keeler, ed., *Sir Francis Drake's West Indian Voyage 1585–86,* Hakluyt Soc. 2d ser. 148 (London: Hakluyt Soc., 1981), 245–46.
16. Guyon's encounter with Redcross and his capture of Furor do not belong in this category. Guyon and Redcross recognize one another before they actually

come to blows (1.26–28), and the Palmer instructs Guyon *not* to use a weapon against Furor (4.10–11).

17. This problem reappears, greatly magnified, in Book 6.
18. Cain sees in the Muslim trappings of Pyrocles and Cymocles "a satiric image of Catholic exploitation of New World peoples egged on by papal sanctions" (98).
19. Las Casas, *Spanish Colonie,* sig. 2r.
20. Spenser, *A View,* 50.
21. René Graziani points out a striking analogy between this episode and the iconography of the Spanish monarchy in the sixteenth century; see his "Philip II's *Impresa* and Spenser's Souldan," *Journal of the Warburg and Courtauld Institutes* 27 (1964): 322–24.
22. Arber, *English Books,* 70.
23. Ibid., 229.
24. See López de Gómara, *Cortés: The Life of the Conqueror by His Secretary,* trans. and ed. Lesley Byrd Simpson (Berkeley: U of California P, 1964), 46; and Bernal Díaz del Castillo, *The Discovery and Conquest of Mexico 1571–1521,* ed. Genaro Gárcia, trans. A. P. Maudslay, intro. Irving A. Leonard (New York: Farrar, 1956), 59. See also Díaz 60–61 for a lengthy anecdote about Cortés's using horses to frighten a group of delegates from the Mayas at Potonchan. A variant of this story appears in Gómara: "Now the horses and mares that were tied to trees in the temple courtyard began to neigh, and the Indians asked Cortés what they were saying. He replied that they were scolding him for not punishing the Indians, at which the Indians offered the horses roses and turkeys to eat, and begged their pardon" (49). Another indication of the importance of horses to the conquistadors is that Díaz includes as part of his history a catalog of the horses that Cortés brought over to Mexico from Cuba, including their names, ownership, appearance, and riding qualities (38–39). Michel de Montaigne, probably borrowing from Gómara, glances at the conquistadors' use of horses in "Des coches": "men mounted on big unknown monsters confronting men who had never seen not merely horses but any animal whatsoever trained to be ridden by man or to bear any other burden" (*The Essays: A Selection,* trans. and ed. M. A. Screech [London: Penguin, 1993], 343). For a dissenting view of the role of the horse, see D. K. Abbass, "Horses and Heroes: The Myth of the Importance of the Horse to the Conquest of the Indies," *Terrae Incognitae: The Journal of the Society for the History of Discoveries* 18 (1986): 21–41. Abbass's examples are highly selective, and the argument conflates events at great chronological distance from each other; ultimately this article does not do much to debunk the long-standing view of the horse as a crucial weapon of the Spanish invasion.
25. Hakluyt, *Principall Navigations,* 2:531.
26. Taylor, *Original Writings,* 2:264.
27. For Cain these stanzas represent "the moral incitement behind a potential imperial realization of Troynovant, which the Virgin Queen had allowed Raleigh to name Virginia" (90–91).
28. Nicholas, *Strange and Delectable History,* sig. ¶v.
29. Taylor, *Original Writings,* 2:256; also quoted in Maltby, *Black Legend in England,* 64.

Chapter 4

1. See also Cain, *Praise,* 95–96.
2. The debate over the significance of Guyon's faint (whether it is physiological, psychological, theological, Christological, moral, mystical, or narratological) was a regular feature of Spenserian scholarship from the 1950s through the 1970s. To summarize that debate very briefly, the main positions were taken by Harry Berger, Frank Kermode, and Paul Alpers. Berger located the meaning of the episode on the narrative level, and concluded that canto 7 is readable in similar terms to moral and psychological fictions of more recent vintage. Kermode detected a neoplatonic mystery concealed within the text. Alpers asserted the precedence of the canto's rhetorical development over its narrative or allegorical development. See Berger, *Allegorical Temper,* 3–38; Kermode, *Shakespeare, Spenser, Donne: Renaissance Essays* (London: Routledge, 1971), 60–83; and Alpers, *The Poetry of "The Faerie Queene"* (1967; reprint, Columbia: U of Missouri P, 1982), 235–75. For modifications of Berger's and Kermode's claims, see Maurice Evans, "The Fall of Guyon," *English Literary History* 28 (1961): 215–24; and Carl Robinson Sonn, "Sir Guyon and the Cave of Mammon," *Studies in English Literature* 1 (1961): 17–30. Most of the relatively small amount of work done on the Cave of Mammon since the 1960s has been by way of sequel to Alpers's critique of Berger's and Kermode's arguments. See Kermode 6–9; Humphrey Tonkin, "Discussing Spenser's Cave of Mammon," *Studies in English Literature* 13 (1973): 1–13; Patrick Cullen, *Infernal Triad: The Flesh, the World and the Devil in Spenser and Milton* (Princeton: Princeton UP, 1974), 68–96; and Roger G. Swearingen, "Guyon's Faint," *Studies in Philology* 74 (1977): 165–85. The introductory pages of Swearingen's essay (165–68) provide a useful overview of the various points of contention over the previous twenty years.
3. There have been few attempts to deal with canto 7 as historical allegory, since the traditional view is that the episode is, in Kermode's words, "without historical dimension" (Kermode, *Renaissance Essays,* 6). An exception is Maureen Quilligan, *Milton's Spenser: The Politics of Reading* (Ithaca, NY: Cornell UP, 1983), 50–72. Quilligan claims that "Spenser's allegory [in 2.7] takes as its province the usually hidden springs of human society, making manifest the latent contradictions of Elizabethan economic organization" (55). Specifically, the Cave of Mammon allegorizes the emerging mercantilism of a financially pressed nobility in the late sixteenth century, and Quilligan refers to the attempts of nobles to mine their own lands in order to bolster income (57). She briefly discusses Ralegh's involving himself in joint-stock colonial and commercial ventures in the New World, "and therefore sharing some of the excess that Spenser censures in the Mammon episode" (68), but her focus is mainly on the English domestic economy. This seems valid as far as it goes—especially given that Ralegh was also Warden of the Stannaries for several years, and thus in charge of the Cornish mines—but it also seems unnecessarily narrow, since the mines that most occupied the attention of the Elizabethans were not on the lands of their own nobles.
4. J. H. Elliott, *The Old World and the New: 1492–1650* (Cambridge: Cambridge UP, 1970), 90.

5. Cain offers a similar, though less specific, interpretation; see Cain, *Praise,* 94.
6. For a reasonably detailed account of the process, see Earl J. Hamilton, *American Treasure and the Price Revolution in Spain, 1501–1650,* Harvard Economic Studies 43 (Cambridge: Harvard UP, 1934), 14–32.
7. As so often happens in the Renaissance, the ancients and moderns turn out to share some common ground, even in the Spanish mines. In his *Discourse of a Discoverie for a new Passage to Cataia* (1576), Sir Humphrey Gilbert cites Marineus Siculus in support of his speculation that the Romans might have reached the New World: "[he] reporteth that there hath been found by the Spaniards, in the Gold Mines of America, certain pieces of Money, engraved with the Image of Augustus Caesar" (*The Voyages and Colonising Enterprises of Humphrey Gilbert,* ed. David Beers Quinn, 2 vols., Hakluyt Soc. 2d ser. 83–84 [London: Hakluyt Soc., 1940] 1:137).
8. Another intriguing feature of these stanzas is Mammon's coat, which, obscured though it is with dust, seems to be "A worke of rich entayle, and curious mould, / Wouen with antickes and wild Imagery" (7.4.5–6). In its curiousness and wildness, this coat suggests not something familiar but something entirely foreign to both the poet's and reader's experience. Specimens of salvaged pre-Colombian art were beginning to trickle into northern Europe in Spenser's time, and accounts of the lavishness and strangeness of Aztec and Inca costumery were widely available in the writings of Cortés, Gómara, Zárate, Oviédo, and Las Casas. For Cain's comment on Mammon and his coat, see *Praise,* 94–95.
9. The likelihood of this increases in light of phrases in Eden's work that appear to be echoed in Book 2. For example, Eden glosses a reference to the Amazon with "the mighty river called Flumen Amazonum, found of late" (Arber, *English Books,* 160); cf. Pr.2.8.
10. One of the books in Dee's library; see Taylor, *Tudor Geography,* 201. There is a modern translation of this work: *The Pirotechnia of Vannoccio Biringuccio,* trans. Cyril Stanley Smith and Martha Teach Gnudi (New York: American Inst. of Mining and Metallurgical Engineers, 1942).
11. Arber, *English Books,* 361.
12. Ibid., 362.
13. Ibid., 355.
14. Nicholas, *Pleasant Historie,* sig. B1r. This is a serious, and fairly typical, misrepresentation of Grijalva, actually a cautious and conscientious explorer who treated the peoples he encountered with relative respect. The negative account of his expedition seems to have originated with Cortés's initial patron (though later his nemesis) Diego Velázquez, the governor of Cuba, who had sent Grijalva out and was disappointed that he returned to Cuba with so little treasure. See William H. Prescott, *History of the Conquest of Mexico and History of the Conquest of Peru* (New York: Modern Library, n.d.), 124–26, and Hugh Thomas, *Conquest: Montezuma, Cortés, and the Fall of Old Mexico* (New York: Touchstone, 1993), 97–115.
15. In Arber's *English Books,* see, for instance, 84, 108, 112–13, 115, 118, 123, 135–36, 140.
16. Arber, *English Books,* 147.
17. Ibid., 107.

18. Ibid., 82.
19. Ibid., 135–36.
20. Ibid., 114.
21. Ibid., 112. Colmenares has the distinction of being the first conquistador to have read out the notorious *Requerimiento,* along the same coastline where Nicuesa's expedition went awry. Oviédo was there to hear him read it. See H. Thomas, *Conquest,* 72.
22. Arber, *English Books,* 117.
23. Ibid.
24. Ibid. For a further analysis of Comogrus's son's speech, see Hadfield, "Peter Martyr, Richard Eden and the New World," 2–12.
25. Arber, *English Books,* 118.
26. Berger, *Allegorical Temper,* 23–26, and Bernhart, "Imperialistic Myth," 352–57.
27. Bernhart does make the observation, based on stanzas 35–37 (Bernhart, "Imperialistic Myth," 347).
28. Arber, *English Books,* 173. Bernhart also cites this passage as it appears in Willes's 1577 edition, but her comparison is to the vines growing on the porch in the Bower of Bliss (Bernhart, "Imperialistic Myth," 288).
29. Arber, *English Books,* 360.
30. Ibid., 149. Most mining in the New World was in fact surface-mining, but the conventional notion of the mine as subterranean prevails even when, as in this passage, a different sort of mine is being discussed. The metaphor of violation here (and incestuous violation at that) might be interestingly compared with Ralegh's much-discussed remark at the end of the *Discoverie of Guiana* that Guiana "is a country that hath yet her maidenhead, never sacked, turned, nor wrought" (Ralegh, *Discoverie of Guiana,* 73). The comparison suggests that Ralegh is not saying anything very striking or original but is actually employing a conventional, paramythological trope applied to underground mining.
31. P. C. Bayley, ed., *The Faerie Queene, Book II* (Oxford: Oxford UP, 1965), 298.
32. Arber, *English Books,* 172. Eden, like other writers of the age, is alert for classical analogies to contemporary situations. At the end of his collection he appends an excerpt from the *Bibliotheca* of Diodorus Siculus titled "The Manner of Working in Gold Mines in Old Time": "The filthiness of the bodies of these laborers, is apparent to all men. For not so much as their privy members are covered with any thing: And their bodies beside so filthy, that no man can behold them without compassion of their misery. But no pity, no rest, no remission is granted them, whether they be men or women, young or old, sick or feeble: But are all with strokes enforced to continual labor until the poor wretches faint and often die for extreme debility" (Arber, *English Books,* 369).
33. Las Casas, *Spanish Colonie,* sig. P2v.
34. Bernhart, "Imperialistic Myth," 347ff. Cain also finds multiple references to the Incas in Book 2; see *Praise,* 92–93, 94–95. Spenser's interest in native American civilization seems rather more generic, however.
35. Nicholas, *Pleasant Historie,* sig. b1.
36. Las Casas, *Spanish Colonie,* sigs. G1v–G2.
37. Berger, *Allegorical Temper,* 29, and Cain, *Praise,* 97–98. See also Berger, *Allegorical Temper,* 22–23.

38. Greenblatt, *Renaissance Self-Fashioning,* 173–74.
39. Taylor, *Original Writings,* 2:246.
40. Ibid., 248.
41. Ibid., 265.
42. A reader of this book has pointed out the intriguing assonance between *Phil*otime and *Phil*ip—as in Philip II. The two could certainly be connected in terms of the symbolic order of this part of canto 7, and Spenser compares Philotime to a male ruler: "neuer earthly Prince in such aray / His glory did enhaunce, and pompous pride display" (7.44.8–9). At the same time she bears comparison to Gloriana the Fairy Queen, which suggests that Philotime could represent a complex, hermaphroditic, perhaps even scurrilously satirical version of the "horns of the dilemma" faced by Guyon here.
43. The conclusion to Guyon's speech—"Then gan a cursed hand the quiet wombe / Of his great Grandmother with steele to wound, / And the hid treasures in her sacred tombe / With Sacriledge to dig" (7.17.1–4)—recalls Martyr's "Spaniard . . . undermining the earth to break the bones of our mother, and enter many miles into her bowels," and Mammon's subsequent response to Guyon ("Thou that doest liue in later times, must wage / Thy workes for wealth, and life for gold engage" [7.18.4–5]) echoes a later part of the same passage from Martyr: "we may now scarcely lead a happy life sith iniquity hath so prevailed and made us slaves to that whereof we are lords by nature." Beyond this possible resonance with the *Decades,* there must be significance in Spenser's use of "Sacriledge" in conjunction with images that carry a certain weight of topicality: defiling a mother's womb, robbing a grave, failing to give proper burial to the dead.

Chapter 5

1. See respectively Pauline Henley, *Spenser in Ireland* (Dublin: Cork UP, 1928), 114–19; Raymond Jenkins, "Spenser and Ireland," *That Soueraine Light: Essays in Honor of Edmund Spenser 1552–1952,* ed. William R. Mueller and Don Cameron Allen (1952; New York: Russell & Russell, 1967), 54; and Clarence Steinberg, "Atin, Pyrocles, Cymocles: On Irish Emblems in 'The Faerie Queene,' " *Neuephilologische Mitteilungen* 72 (1971): 749–61.
2. Bayley, *The Faerie Queene: Book II,* 304.
3. M. M. Gray, "The Influence of Spenser's Irish Experiences on *The Faerie Queene,*" *Review of English Studies* 6 (1930): 415–16; for Henley, see n. 1 above. Gray claims that the opening tableau of canto 11 "owes something to the Munster rebellion" (416); this makes little sense from a historical standpoint, since the events in Munster—and those leading to Spenser's personal calamity—occurred roughly eight years after Book 2's completion. Gray's article prompted a gracious refutation from C. S. Lewis; see *Review of English Studies* 7 (1931): 83–85. Lewis's letter cites instead the Italianate influences (especially from Boiardo) on Book 2.
4. Greenblatt, *Renaissance Self-Fashioning,* 186.
5. Gray, "Spenser's Irish Experiences," 415.
6. Cain notes the analogy with poisoned arrows (Cain, *Praise,* 99). Linton, in *Romance* (91), notes the New World provenance of Maleger's arrows and suggests a similar provenance for his Antaeus-like powers. Hamlin also discusses

the probability of Maleger's native American lineage, based on both the arrows and the tiger which serves as Maleger's mount; on the tiger, see my discussion below (Hamlin, *Images of America,* 72–73).

7. Gray, "Spenser's Irish Experiences," 414.
8. Spenser, *A View,* 52.
9. Ibid., 51.
10. Ibid., 52.
11. That Maleger wears a "dead mans skull" for a helmet is, among other allegorical possibilities, an association with cannibalism and human sacrifice. Skulls both real and artificial were used ornamentally in Aztec culture. The Aztecs, of course, wore cotton armor.
12. Arber, *English Books,* 113. This king might be compared as well to the deceptively feminine figure of Genius in the Bower of Bliss.
13. Moreover, the Indians' tactics approximate Maleger's—"Apace he shot, and yet he fled apace" (11.27.1)—as suggested in Martyr's description of Martín Fernández de Enciso's ambush by three cannibals "armed with bows and venomous arrows, who without all fear, assailed our men fiercely, wounded many, and slew many: And when they had emptied their quivers, fled as swiftly as the wind: For . . . they are exceeding swift of foot by reason of their loose going from their childs age. They affirm that they let slip no arrow out of their bows in vain" (Arber, *English Books,* 110). Spenser compares the speed of Maleger's fleeing tiger to "the winged wind" (11.26.1).
14. Spenser, *A View,* 57.
15. There is classical precedent for the claim that the Scythians also used poisoned arrows as weapons. Lucan speaks of this in the *Pharsalia:* "Tinxere sagittas / Errantes Scythiae populi" (3.266–67). But while Spenser makes a connection between the Scythians and the Irish in *A View,* no similar connection is evident in Book 2 of *The Faerie Queene.*
16. The *Oxford English Dictionary*'s definition of "panther" includes an interesting quotation from an 1894 issue of *Century Magazine,* which serves to point out that old ideas die slow deaths: "The panther was long called a 'tyger' in the Carolinas, and a 'lyon' elsewhere."
17. Arber, *English Books,* 144.
18. Ibid., 235.
19. Ibid., 236.
20. Ibid., 52.
21. Ibid., 178.
22. A. S. P. Woodhouse, "Nature and Grace in *The Faerie Queene,*" *Essential Articles for the Study of Edmund Spenser,* ed. A. C. Hamilton (Hamden, CT: Archon, 1972), 78.
23. Las Casas, *Spanish Colonie,* sig. A1v.
24. Ibid., sig. A1r.
25. Taylor, *Original Writings,* 2:214.
26. Establishing the actual pre-Columbian population of the New World has proven to be extremely difficult, and controversy has dogged the various efforts to do so. In his brief but useful discussion of population estimates for Mexico, Hugh Thomas concludes, "All that can really be said of these last fifty years of research

and speculation is that there have been many brilliant calculations but that, in the end, nobody can be shown to have made anything more than an inspired guess" (613). See H. Thomas, *Conquest,* 609–14.

27. Nicholas, *Pleasant Historie,* 264.
28. Ibid., 265.
29. Ibid., 277.
30. Cain mentions another possible explanation for Maleger's terrible pertinacity: his name, his appearance, and the fact that he is accompanied by "two wicked Hags" named Impotence and Impatience (11.23.2, 8, 9) could suggest that Maleger symbolizes—and may even be suffering from—the ravages of venereal disease. Syphilis was reputed to be one of earliest exports from the New World to the Old. In the sixteenth century, at least, it was both epidemic and incurable. Cain draws his inferences from Francisco Guerra, "The Problem of Syphilis," Chiappelli 2:845–51, a compact account of the subject and the issues it raises for New World history (Cain, *Praise,* 99).
31. See, for instance, Hadfield, *Spenser's Irish Experience,* 163–64.
32. In addition to the works cited in the introduction, see A. L. Rowse, *The Expansion of Elizabethan England* (New York: St. Martin's, 1955), 136–37; David Beers Quinn, *The Elizabethans and the Irish* (Ithaca, NY: Cornell UP, 1966), 106–22; Nicholas P. Canny, "The Ideology of English Colonization: From Ireland to America," *William and Mary Quarterly,* 3d ser., 30 (1973): 575–98 and *The Elizabethan Conquest of Ireland: A Pattern Established 1565–76* (New York: Barnes, 1976), 160–63; and Karl S. Bottigheimer, "Kingdom and Colony: Ireland in the Westward Enterprise 1536–1660," *The Westward Enterprise: English Activity in Ireland, the Atlantic and America 1480–1650,* ed. K. R. Andrews, N. P. Canny, and P. E. H. Hair (Detroit: Wayne State UP, 1979), 45–64.
33. In his commentary on *A View,* Renwick includes a passage of text that only appears in the Public Record House manuscript, showing that Spenser had at least passing thoughts about New World matters during the time he was writing about Ireland. The passage (also quoted by S. Miller [*Invested with Meaning,* 71] to support an "Irish" thesis quite different from my own) contains Spenser's doubts about Ireland having been systematically peopled by the Spanish: "if there were any such notable transmission of a Colony hither out of Spain, or any such famous Conquest of this kingdom by . . . a Spaniard . . . it is not unlikely but that the very Chronicles of Spain, had Spain then been in so high regard . . . would not have omitted so memorable a thing, as the subduing of so noble a Realm, to the Spaniard, no more than they do now neglect to memorize their Conquest of the Indians" (197). Spenser's concern at the end of this passage is with a literary question: he refers not to the actual events of conquest but to their subsequent narrations—indeed the only points of contact that most English citizens had with the events that had transpired across the Atlantic.

Chapter 6

1. Elliott, *Old World,* 21.
2. Arber, *English Books,* 112. Enciso was accompanied on this expedition by Balboa—then a stowaway, trying to escape from his creditors. A much more forceful character than Enciso, Balboa eventually seized control of the colony

and expelled Enciso entirely. It was after his return to Spain that Enciso became famous as a geographer. See Penrose, *Travel and Discovery,* 119–20, 365–66.

3. Arber, *English Books,* 163.
4. Ibid., 73, 165, 176.
5. Occasionally, Martyr compares data from the new discoveries with the claims of old folklore, so as to arrive at a rational explanation of the latter in view of the former. He makes sense of sirens and Amazons, for instance, by resorting to a bit of rough-hewn anthropological observation involving the Caribs: "When the men go forth of the land a man hunting, the women manfully defend their coasts against such as attempt to invade the same. And hereby I suppose it was thought that there were Islands in the Ocean, inhabited only with women" (Arber, *English Books,* 177).
6. Spenser's strategy resembles Martyr's quite closely at stanzas 11–13, where the Boatman makes sense of the Wandering Islands in terms of a classical model: "As th'Isle of *Delos* whylome men report / Amid th'*Aegæan* sea long time did stray" (12.13.1–2).
7. A phenomenon noted in Las Casas's redaction of Columbus's logbook, in the entry for September 23, 1492: "As the sea was calm and smooth the crew grumbled, saying that since there were no heavy seas in these parts no wind would ever blow to carry them back to Spain. But later the seas rose high without any wind and this astonished them. The Admiral says at this point: 'I was in great need of these high seas because nothing like this had occurred since the time of the Jews when the Egyptians came out against Moses who was leading them out of captivity' " (Cohen, *Christopher Columbus,* 45).
8. Arber, *English Books,* 170.
9. Cain suggests that Spenser reveals an awareness of the New World voyage literature by including this catalog of sea monsters, which has no analog in canto 12's ostensible main literary source, the Armida episode of Tasso's *Gerusalemme Liberata.* Cain does not, however, demonstrate that this catalog exists in such a pure form anywhere within the voyage literature (Cain, *Praise,* 91–92).
10. Arber, *English Books,* 103.
11. Ibid., 108.
12. A. C. Hamilton, *The Structure of Allegory in "The Faerie Queene"* (Oxford: Oxford UP, 1961), 12.
13. See, most recently, S. Miller, *Invested with Meaning,* 43–49.
14. Lewis's ideas on *The Faerie Queene* are further developed in N. S. Brooke's essay "C. S. Lewis and Spenser: Nature, Art and the Bower of Bliss," in which the Bower is portrayed as a realm of disorder which, implicitly, must be reordered by Guyon. Brooke's essay and the relevant material from *The Allegory of Love* appear in *Essential Articles for the Study of Edmund Spenser,* ed. A. C. Hamilton (Hamden, CT: Archon, 1972), 3–28.
15. On this point see David Lee Miller, *The Poem's Two Bodies: The Poetics of the 1590 "Faerie Queene"* (Princeton: Princeton UP, 1988), 286–87.
16. See Cain, *Praise,* 92–93, and Bernhart, "Imperialistic Myth," 288, 290. Both draw their information on such gardens from Ralegh's *Discoverie of Guiana.* Bernhart states, "Acrasia's garden paradise of luxurious artificiality is clearly related to contemporary reports of the riches of Spain's overseas dominions"

(286). It is worth noting, however, that "gardens" of this sort were purported to exist in the Old World as well. Eden's selections from Biringuccio contain an account of such a garden in Hungary: "pure gold springeth out of the earth in the likeness of small herbs, wreathed and twined like small stalks of hops, about the bigness of a pack thread, and four fingers in length or some a handful. . . . The which (if it be true) surely the husband men of these fields shall reap heavenly and not earthly fruits, sent them of god from heaven, and brought forth of nature without their travail or art" (Arber, *English Books,* 364).

17. Zárate, *Discouerie and Conquest,* sig. C3v. This little island was most likely Santa Clara. Prescott describes it as follows: "The place was uninhabited, but was recognized by the Indians . . . as occasionally resorted to by the warlike people of the neighboring isle of Puná, for purposes of sacrifice and worship. The Spaniards found on the spot a few bits of gold rudely wrought into various shapes, and probably designed as offerings to the Indian deity" (Prescott, *History of the Conquest of Peru,* 869). What struck sixteenth-century observers about Incan handiwork in precious metals was not so much its complex artifice as the way in which gold and silver were made to serve the most humble purposes. In his remarks on Peru, Gómara notes that "gold is there in such plenty that they make pisspots thereof, and other vessels applied to filthy uses. But this is more to be marvelled at, that in a city called Collao was found a house all covered with massy plates of gold" (Arber, *English Books,* 343). Perhaps the most marvellous thing about the house from Gómara's point of view is that the Incas have transformed gold into a common building material.
18. See 16.9–10 in Torquato Tasso, *Jerusalem Delivered,* trans. Edward Fairfax, intro. John Charles Nelson (New York: Capricorn, n.d.), 320; Armida's garden is, of course, also a work of art (16.10.1–4).
19. Arber, *English Books,* 87.
20. Ibid., 167–68. In Eden's excerpts from Oviédo, under the heading "Of trees, fruits, and plants," there is an interesting reflection on the limitations of ancient geography as a point of reference for interpreting the New World. Oviédo observes that Pliny "nameth not past five or six" evergreen trees in his natural history. However, "in the Islands of these Indies, and also in the firm land, it is a thing of much difficulties to find two trees that lose or cast their leaves at any time" (Arber, *English Books,* 227).
21. Ibid., 76.
22. Ibid., 89.
23. Ibid., 83.
24. Perhaps the most flagrant instance of Fairfax's borrowing is his rendering at 15.62.1–2 of Spenser's "Withall she laughed, and she blusht withall, / That blushing to her laughter gaue more grace, / And laughter to her blushing" (12.68.1–3) as "Withal she smiled, and she blushed withal, / Her blush her smiling, smiles her blushing graced" (Tasso, *Jerusalem Delivered,* 316).
25. Tasso, *Jerusalem Delivered,* 309. On the meaning of the Columbus passage within the epic tradition, see Quint, *Epic and Empire,* 248–53.
26. Greenblatt, *Renaissance Self-Fashioning,* 175.
27. Arber, *English Books,* 37.
28. It may be worth recalling here the beginnings of Guyon's quest in the tragedy

of Mordant and Amavia, which suggests the antithesis of painless childbirth. Ruddymane's "birth" is accompanied by the violent death of both his parents and signalized by his irremediably bloody hands. For further discussion see chapter 2.

29. Arber, *English Books,* 67.
30. Ibid., 69.
31. Ibid., 190. The quotation in *Renaissance Self-Fashioning* appears on 181 and is slightly different in wording, as it comes from a nineteenth-century translation of Martyr.
32. Ibid., 156.
33. Ibid., 138. On the subject of berdache culture in the Americas, see Richard C. Trexler's genuinely eye-opening study *Sex and Conquest: Gendered Violence, Political Order, and the European Conquest of the Americas* (Ithaca, NY: Cornell UP, 1995), esp. 82–101.
34. The passage does not make clear if the offending brother (or the king himself) was among Balboa's victims. The text seems to indicate that Balboa chose to punish only the consorts.
35. Is it implausible to read into Genius's ability to "entize" travelers the suggestion of a prostitute's stratagems?
36. Cf. Martyr on the swine of Hispaniola: "The multitude of hogs, are exceedingly increased, and become wild as soon as they are out of the swineherds keeping" (Arber, *English Books,* 106). Spenser's obvious debt in canto 12 to the Circe episode of the *Odyssey* ought not to obscure his sensitivity to the patriotic dimensions of Homer's narrative. That Odysseus's men unwittingly allow themselves to be transformed by Circe into swine can be construed not only as a weakness in these particular men but also as a flaw in the Ithakans' national character: they have forgotten their homeland in their anxiety to participate in "unmanly" pleasures. Odysseus restores the balance, but he is also the only Ithakan to survive the journey home.
37. This corrective act can also be regarded as one of conversion, such as occurs in the following passage from the *Decades* concerning the Christianizing of the king of the island of Margarita: "[The Spaniards] . . . converted him from a cruel tiger to one of the meek sheep of Christs flock sanctified with the water of baptism with all his family and kingdom" (Arber, *English Books,* 178). Strikingly, Eden glosses this passage with the remark that "Wild beasts must be tamed with the rod." Given the context, Eden appears to be referring to "the rod" in a typological sense appropriate to the event (i.e., as Aaron's rod/shepherd's crook/crucifix) and not as an instrument of punishment. Such taming is close to what the Palmer does at stanza 40 with his "mighty staffe," so much like "the rod of *Mercury*" (12.40.3, 41.3), and again at stanza 86 when he converts the beasts to "comely men" (12.86.2).
38. See, for instance, Arber, *English Books,* 70, 104, 138, 148, 172.
39. Arber, *English Books,* 76.
40. Ibid., 79.
41. Ibid., 80. According to Columbus himself, the lassitude of his men eventually bordered on absurdity, the weight of which the natives literally had to bear. The Spaniards were "so given to Idleness and sleep, that whereas they were brought

thither for miners, laborers, and scullions, they would not now go one furlong from their houses except they were borne on other men's backs . . . For, to this office, they put the miserable Island men, whom they handled most cruelly" (Arber, *English Books,* 91).

42. Zárate, *Discouerie and Conquest,* sig. B2r–v.
43. Elliott remarks that the problem was observable within Spain's borders as well as outside them, for instance in Seville, the point of entry for much of the New world treasure. "The city acted as a magnet for the population of Castile—for the restless, the ambitious and the hungry, who drifted southwards in the hope of sharing, at least at secondhand, in the prosperity brought by the Indies. The splendours and miseries of the teeming streets of Seville—a city of 100,000 or more by the end of the sixteenth century—provided the most striking visual evidence anywhere in Europe of the impact of America on sixteenth-century life" (Elliott, *Old World,* 75).
44. Spenser stresses this equivalence again with the removal of the captives from the Bower: "Then led they her away, *and eke* that knight / They with them led, *both* sorrowfull and sad" (12.84.1–2; my emphases).
45. Is there even the possibility of a topical weight to these "captiue bandes," given the ubiquity of slavery and enforced labor in colonial economies?
46. The works of Lewis Hanke offer the best English-language introduction to the complex history of clerical-secular relations in the Spanish New World; see especially Hanke, *The Spanish Struggle for Justice in the Conquest of America* (Philadelphia: U of Pennsylvania P, 1949).
47. Arber, *English Books,* 50.
48. For a more or less "Irish" interpretation of the Palmer's parting words, see S. Miller, *Invested with Meaning,* 47–49.
49. Sidney, *Defence of Poetry,* 52.
50. Ibid., 53.

Conclusion

1. One could argue that Ralegh's two voyages to Guiana are a tragicomic enactment of this very crisis. Several recent studies of Ralegh touch on this notion. See Nicholl, *Creature in the Map;* Fuller, *Voyages in Print,* 55–84; and Louis Montrose, "The Work of Gender in the Discourse of Discovery," Stephen Greenblatt, ed., *New World Encounters* (Berkeley: U of California P, 1993), 177–217.
2. *1 Henry IV* 4.3.66–67, cited from *The Riverside Shakespeare,* 2d ed. (Boston: Houghton, 1997).
3. It should be noted that on his first Guiana expedition Ralegh named a branch of the Orinoco the "Redcross" river, but the name did not stick. On echoes of the 1590 edition of *The Faerie Queene* in the *Discoverie of Guiana,* see Nicholl, *Creature in the Map,* 304–13; Mackenthun, *Metaphors of Dispossession,* 173–74; and Montrose, "Work of Gender," 187, 197, 209.
4. See Louis Althusser, *Lenin and Philosophy and Other Essays,* trans. Ben Brewster (New York: Monthly Review, 1971), 127–86.
5. Helgerson, *Forms of Nationhood,* 13. Helgerson says two sentences earlier that "texts and nations . . . are also agents, making things happen in the world of men and women and the face-to-face groups of which men and women are

part" (13). This statement depends on a very generalized notion of agency; groups, individuals, nations, and texts do not "make things happen," or "face" each other, in exactly the same way.

6. S. Miller uses a similar riverine metaphor to describe "influence" (Miller, *Invested with Meaning,* 1–2).
7. A. C. Hamilton, ed., *The Faerie Queene,* 737.
8. Ludwig Wittgenstein, *Tractatus Logico-Philosophicus,* trans. D. F. Pears and B. F. McGuinness, intro. Bertrand Russell (London: Routledge, 1974), 5.
9. Greenblatt, *Renaissance Self-Fashioning,* 5.
10. Brook Thomas, *The New Historicism and Other Old-Fashioned Topics* (Princeton: Princeton UP, 1991), 196.
11. B. Thomas, *New Historicism,* 201, 203.
12. "Testimony" in the sense of a discourse that is both personal and public, but not necessarily authoritative.
13. B. Thomas, *New Historicism,* 215.
14. Gerald L. Bruns, *Hermeneutics Ancient and Modern* (New Haven: Yale UP, 1992), 184.
15. Bruns, *Hermeneutics,* 217. Interestingly, Bruns designates allegory as the mode in which the "condition of exposure" is least likely to be experienced: "Allegory is a mode of translation that rewrites an alien discourse in order to make it come out right—according to prevailing norms of what is right. . . . [It] is appropriative discourse with implicit claims to universality in the sense that there is, theoretically, nothing that it is required to reject as alien or just plain false. It enlarges its empire of learning by taking the strangeness at its borders as a difference of letter and spirit" (Bruns, *Hermeneutics,* 203). For Bruns, the antithesis to allegory, the optimally "exposing" mode, is satire. This account of allegory has explanatory power in a general way; however, it does not seem to make adequate sense of Spenser's allegory, which is certainly "appropriative" but retains its "strangeness" in spite of its gestures toward "universality." *The Faerie Queene* offers many of the effects of satire (at least as Bruns invokes that term) without presenting satire's conventional generic markers. Within the canon of allegory, then, Spenser may well be the exception who proves the rule—and such a distinction may help to account for the fact that his work has become relatively marginalized in the academy, since it never quite generates the ideological coherence to be described as "normative" or "representative" in the sense that critics like to apply these adjectives.
16. B. Thomas, *New Historicism,* 215.
17. See the letter to Ralegh, A. C. Hamilton, ed., *The Faerie Queene,* 737.

Works Cited

References in the text to individual entries in *The Spenser Encyclopedia* are not listed separately.

Abbass, D. K. "Horses and Heroes: The Myth of the Importance of the Horse to the Conquest of the Indies." *Terrae Incognitae: The Journal of the Society for the History of Discoveries* 18 (1986): 21–41.

Alpers, Paul. *The Poetry of "The Faerie Queene."* 1967. Reprint, Columbia: U of Missouri P, 1982.

Althusser, Louis. *Lenin and Philosophy and Other Essays.* Trans. Ben Brewster. New York: Monthly Review, 1971.

Anderson, Judith H. "The Knight and the Palmer in *The Faerie Queene.*" *Modern Language Quarterly* 31 (1970): 160–78.

——— et al., eds. *Spenser's Life and the Subject of Biography.* Amherst: U of Massachusetts P, 1996.

Andrews, Kenneth R. *Trade, Plunder and Settlement: Maritime Enterprise and the Genesis of the British Empire 1480–1630.* Cambridge: Cambridge UP, 1984.

——— et al., eds. *The Westward Enterprise: English Activity in Ireland, the Atlantic and America 1480–1650.* Detroit: Wayne State UP, 1979.

Arber, Edward, ed. *The First Three English Books on America. [1511]–1555 A.D.* Westminster, 1895. (Reprints text of Martyr's *The Decades of the newe worlde.*)

Auden, W. H. *Selected Poetry of W. H. Auden.* 2d ed. New York: Random-Vintage, 1971.

Baker, David J. *Between Nations: Shakespeare, Spenser, Marvell, and the Question of Britain.* Stanford: Stanford UP, 1997.

Berger, Harry, Jr. *The Allegorical Temper: Vision and Reality in Book II of Spenser's "Faerie Queene"*. Yale Studies in English 137. New Haven: Yale UP, 1957.
Berleth, Richard. *The Twilight Lords: An Irish Chronicle.* New York: Knopf, 1978.
Bernhart, Barbara-Maria. "Imperialistic Myth and Iconography in Books I and II of the Faerie Queene." Ph.D. diss., McMaster University, 1975.
Biringuccio, Vannoccio. *The Pirotechnia of Vannoccio Biringuccio.* Trans. Cyril Stanley Smith and Martha Teach Gnudi. New York: American Inst. of Mining and Metallurgical Engineers, 1942.
Bottigheimer, Karl S. "Kingdom and Colony: Ireland in the Westward Enterprise 1536–1660." In Andrews et al., *Westward Enterprise,* 45–64.
Bradshaw, Brendan, Andrew Hadfield, and Willy Maley, eds. *Representing Ireland: Literature and the Origins of Conflict, 1534–1660.* Cambridge: Cambridge UP, 1993.
Brink, Jean R. "'All his minde on honour fixed': The Preferment of Edmund Spenser." In Anderson et al., *Spenser's Life,* 45–64.
———. "Constructing the *View of the Present State of Ireland.*" *Spenser Studies: A Renaissance Poetry Annual* 11 (1990): 203–28.
Brooke, N. S. "C. S. Lewis and Spenser: Nature, Art and the Bower of Bliss." In A. C. Hamilton, *Essential Articles,* 13–28.
Bruns, Gerald L. *Hermeneutics Ancient and Modern.* New Haven: Yale UP, 1992.
Cain, Thomas H. *Praise in "The Faerie Queene."* Lincoln: U of Nebraska P, 1978.
Campbell, Mary B. *The Witness and the Other World: Exotic European Travel Writing, 400–1600.* Ithaca, NY: Cornell UP, 1988.
Canny, Nicholas P. *The Elizabethan Conquest of Ireland: A Pattern Established 1565–76.* New York: Barnes, 1976.
———. "The Ideology of English Colonization: From England to America." *William and Mary Quarterly* 3d ser. 30 (1973): 575–98.
Carey, Vincent P., and Clare L. Carroll. "Factions and Fictions: Spenser's Reflections of and on Elizabethan Politics." In Anderson et al., *Spenser's Life,* 31–44.
Cawley, Robert Ralston. *Unpathed Waters: Studies in the Influence of the Voyagers on Elizabethan Literature.* Princeton: Princeton UP, 1940.
Cheyfitz, Eric. *The Poetics of Imperialism: Translation and Colonization from "The Tempest" to "Tarzan."* New York: Oxford UP, 1991.
Chiappelli, Fredi, ed. *First Images of America: The Impact of the New World on the Old.* 2 vols. Berkeley: U of California P, 1976.
Clulee, Nicholas. *John Dee's Natural Philosophy: Between Science and Religion.* London: Routledge, 1988.
Cohen, J. M., ed. and trans. *The Four Voyages of Christopher Columbus.* Harmondsworth, England: Penguin, 1969.
Collinson, Richard, ed. *The Three Voyages of Martin Frobisher.* Hakluyt Soc. 38. London, 1867.
Cullen, Patrick. *Infernal Triad: The Flesh, the World and the Devil in Spenser and Milton.* Princeton: Princeton UP, 1974.
Diaz del Castillo, Bernal. *The Discovery and Conquest of Mexico 1517–1521.* Trans. A. P. Maudslay. Intro. Irving A. Leonard. New York: Farrar, 1956.
Dudley, Edward, and Maximilian E. Novak, eds. *The Wild Man Within: An Image*

in Western Thought from the Renaissance to Romanticism. Pittsburgh: U of Pittsburgh P, 1972.

Elliott, J. H. *The Old World and the New: 1492–1650.* Cambridge: Cambridge UP, 1970.

Evans, J. Martin. *Milton's Imperial Epic: Paradise Lost and the Discourse of Colonialism.* Ithaca, NY: Cornell UP, 1996.

Evans, Maurice. "The Fall of Guyon." *English Literary History* 28 (1961): 215–24.

Erickson, Wayne. *Mapping "The Faerie Queene": Quest Structures and the World of the Poem.* New York: Garland, 1996.

Farness, Jay. "Disenchanted Elves: Biography in the Text of *Faerie Queene* V." In Anderson et al., *Spenser's Life,* 18–30.

Fowler, A. D. S. "Emblems of Temperance in *The Faerie Queene,* Book II." *Review of English Studies* n.s. 11 (1961): 143–49.

French, Peter. *John Dee: The World of an Elizabethan Magus.* 1972. Reprint, London: Ark-Routledge, 1987.

Friede, Juan, and Benjamin Keen, eds. *Bartolomé de las Casas in History: Toward an Understanding of the Man and His Work.* De Kalb: Northern Illinois UP, 1971.

Frushell, Richard C., and Bernard J. Vondersmith, eds. *Contemporary Thought on Edmund Spenser: With a Bibliography of Criticism of "The Faerie Queene," 1900–1970.* Carbondale: Southern Illinois UP, 1975.

Fuller, Mary C. *Voyages in Print: English Travel to America, 1576–1624.* Cambridge Studies in Renaissance Literature and Culture 7. Cambridge: Cambridge UP, 1995.

Garber, Marjorie, ed. *Cannibals, Witches, and Divorce: Estranging the Renaissance.* Selected Papers from the English Institute, n.s. 11 (1985). Baltimore: Johns Hopkins UP, 1987.

Giamatti, A. Bartlett. "Primitivism and the Process of Civility in Spenser's *Faerie Queene.*" In Chiappelli, *First Images of America,* 1:71–82.

Gibson, Charles, ed. *The Black Legend: Anti-Spanish Attitudes in the Old World and the New.* New York: Knopf, 1971.

Gillies, John. *Shakespeare and the Geography of Difference.* Cambridge Studies in Renaissance Literature and Culture 4. Cambridge: Cambridge UP, 1994.

Gray, M. M. "The Influence of Spenser's Irish Experiences on *The Faerie Queene.*" *Review of English Studies* 6 (1930): 415–16.

Graziani, René. "Philip II's *Impresa* and Spenser's Souldan." *Journal of the Warburg and Courtauld Institutes* 27 (1964): 322–24.

Greenblatt, Stephen. "Learning to Curse: Aspects of Linguistic Colonialism in the Sixteenth Century." In Chiappelli, *First Images of America,* 2:561–80

———. *Learning to Curse: Essays in Early Modern Culture.* New York: Routledge, 1990.

———. *Marvellous Possessions: The Wonder of the New World.* Chicago: U of Chicago P, 1991.

———, ed. *New World Encounters.* Berkeley: U of California P, 1993.

———. *Renaissance Self-Fashioning: From More to Shakespeare.* Chicago: U of Chicago P, 1980.

———. *Shakespearean Negotiations: The Circulation of Social Energy in Renaissance England.* Berkeley: U of California P, 1988.

———. *Sir Walter Ralegh: The Renaissance Man and His Roles.* New Haven: Yale UP, 1973.

Greenlaw, Edwin. "Spenser and British Imperialism." *Modern Philology* 9 (1912): 347–70.

Greville, Fulke, *Life of Sir Philip Sidney.* Oxford: Clarendon, 1907.

Hadfield, Andrew. *Edmund Spenser's Irish Experience: Wilde Fruit and Salvage Soyl.* Oxford: Clarendon, 1997.

———. "Peter Martyr, Richard Eden and the New World: Reading, Experience, and Translation." *Connotations* 5 (1995–96): 1–22.

———, and Willy Maley. "Introduction: Irish Representations and English Alternatives." In Bradshaw et al., *Representing Ireland,* 1–23.

Hakluyt, Richard. *The Principall Navigations, Voiages and Discoveries of the English Nation.* Hakluyt Soc. Facsimile ed. 2 vols. Cambridge: Cambridge UP, 1965.

Hamilton, A. C., ed. *Essential Articles for the Study of Edmund Spenser.* Hamden, CT: Archon, 1972.

———, ed. *The Faerie Queene.* Longman Annotated English Poets Series. London: Longman, 1977.

———, ed. *The Spenser Encyclopedia.* Toronto: U of Toronto P, 1990.

———. *The Structure of Allegory in "The Faerie Queene".* Oxford: Oxford UP, 1961.

Hamilton, Earl J. *American Treasure and the Price Revolution in Spain, 1501–1650.* Harvard Economic Studies 43. Cambridge: Harvard UP, 1934.

Hamlin, William M. *The Image of America in Montaigne, Spenser, and Shakespeare: Renaissance Ethnography and Literary Reflection.* New York: St. Martin's, 1995.

Hampden, John, ed. *Francis Drake Privateer: Contemporary Narratives and Documents.* London: Eyre Methuen, 1972.

Hanke, Lewis. *The Spanish Struggle for Justice in the Conquest of America.* Philadelphia: U of Pennsylvania P, 1949.

Hardison, O. B., Jr., ed. *English Literary Criticism: The Renaissance.* New York: Goldentree-Appleton, 1963.

Har[r]iot, Thomas. *A Briefe and True Report of the New Found Land of Virginia: The Complete 1590 Theodore de Bry Edition.* Intro. Paul Hulton. New York: Dover, 1972.

Helgerson, Richard. *Forms of Nationhood: The Elizabethan Writing of England.* Chicago: U of Chicago P, 1992.

Hemming, John. *The Search for El Dorado.* New York: Dutton, 1978.

Henley, Pauline. *Spenser in Ireland.* Dublin: Cork UP, 1928.

Highley, Christopher. *Shakespeare, Spenser, and the Crisis in Ireland.* Cambridge Studies in Renaissance Literature and Culture 23. Cambridge: Cambridge UP, 1997.

Hulme, Peter. *Colonial Encounters: Europe and the Native Caribbean, 1492–1797.* London: Methuen, 1986.

Hume, Martin. *Spanish Influence on English Literature.* 1904. Reprint, New York: Haskell House, 1964.

Jardine, Lisa. "Encountering Ireland: Gabriel Harvey, Edmund Spenser, and English Colonial Ventures." In Bradshaw et al., *Representing Ireland,* 60–75.

Jenkins, Raymond. "Spenser and Ireland." In Mueller and Allen, *That Soueraine Light,* 51–62.

Jones, Howard Mumford. *O Strange New World: American Culture: The Formative Years.* New York: Viking, 1964.

Judson, Alexander C. *The Life of Edmund Spenser.* Baltimore: Johns Hopkins UP, 1945.

Keeler, Mary Frear, ed. *Sir Francis Drake's West Indian Voyage 1585–86.* Hakluyt Soc. 2d ser. 148. London: Hakluyt Soc., 1981.

Kermode, Frank. *Shakespeare, Spenser, Donne: Renaissance Essays.* London: Routledge, 1971.

Knapp, Jeffrey. *An Empire Nowhere: England, America, and Literature from "Utopia" to "The Tempest."* The New Historicism: Studies in Cultural Poetics 16. Berkeley: U of California P, 1992.

Las Casas, Bartolomé. *History of the Indies.* Trans. and ed. Andrée Collard. New York: Harper and Row, 1971.

———. *In Defense of the Indians: The Defense of . . . las Casas . . . Against the Persecutors and Slanderers of the Peoples of the New World Discovered Across the Seas.* Trans. Stafford Poole. DeKalb: Northern Illinois UP, 1974.

———. *The Spanish Colonie, or Briefe Chronicle of the Acts and gests of the Spaniardes in the West Indies, called the newe World, for the space of xl. yeares: written in the Castilian by the reverend Bishop Bartholomew de las Casas or Casaus, a Friar of the order of S. Dominicke. And nowe first translated into english, by M. M. S. Imprinted at London [by Thomas Dawson] for William Bro[o]me. 1583.* (Translation of *Brevissima Relación de la destruycion de las Indias.*)

Lestringant, Frank. *Cannibals: The Discovery and Representation of the Cannibal from Columbus to Jules Verne.* Trans. Rosemary Morris. The New Historicism: Studies in Cultural Poetics 37. Berkeley: U of California P, 1997.

———. *Mapping the Renaissance World: The Geographical Imagination in the Age of Discovery.* Trans. David Fausett. Intro. Stephen Greenblatt. The New Historicism: Studies in Cultural Poetics 32. Berkeley: U of California P, 1994.

Lewis, C. S. *The Allegory of Love.* Oxford: Clarendon, 1936.

———. *English Literature in the Sixteenth Century Excluding Drama,* Oxford History of English Literature 3. Oxford: Oxford UP, 1954.

———. Letter to the editor. *Review of English Studies* 7 (1931): 83–85.

Lim, Walter S. H. *The Arts of Empire: The Poetics of Colonialism from Ralegh to Milton.* Newark: U of Delaware P, 1998.

Linton, Joan Pong. *The Romance of the New World: Gender and the Literary Formations of English Colonialism.* Cambridge Studies in Renaissance Literature and Culture 27. Cambridge: Cambridge UP, 1998.

Lopez de Gómara, Francisco. *Cortés: The Life of the Conqueror by His Secretary.* Trans. and ed. Lesley Byrd Simpson. Berkeley: U of California P, 1964.

Lupton, Julia Reinhard. "Mapping Mutability: Or, Spenser's Irish Plot." In Bradshaw et al., *Representing Ireland,* 93–115.

Mackenthun, Gesa. *Metaphors of Dispossession: American Beginnings and the Translation of Empire, 1492–1637.* Norman: U of Oklahoma P, 1997.

Maley, Willy. *Salvaging Spenser: Colonialism, Culture and Identity.* New York: St. Martin's, 1997.

Maltby, William S. *The Black Legend in England: The Development of Anti-Spanish Sentiment, 1558–1660.* Durham, NC: Duke UP, 1971.

Markham, Clements R., trans. *The Letters of Amerigo Vespucci and Other Documents Illustrative of His Career.* Hakluyt Soc. 90. London: Hakluyt Soc., 1894.

Martyr, Peter. *The Decades of the newe worlde or west India, Conteyning the nauigations and conquestes of the Spanyardes, with the particular description of the moste ryche and large landes and landes lately founde in the west Ocean perteyning to the inheritaunce of the kinges of Spayne. In which the diligent reader may not only consyder what commoditie may hereby chaunce to the hole christian world in tyme to come, but also learne many secreates touchynge the lande, the sea, and the starres, very necessarie to be knowen to al such as shal attempte any nauigations, or otherwise haue delite to beholde the strange and woonderfull woorkes of God and nature. Wrytten in the Latine tounge by Peter Martyr of Angleria, and translated into Eglysshe by Rycharde Eden. Londini. In aedibus Guilhelmi Powell. Anno. 1555.* (Reprinted in Arber; also known as *Decades de Orbe Novo.*)

McCabe, Richard A. "Edmund Spenser, Poet of Exile." *Proceedings of the British Academy* 80 (1993): 73–103.

McKerrow, Ronald B., ed. *The Works of Thomas Nashe Edited from the Original Texts.* Vol. 1. London: Sidgwick & Jackson, 1910.

Miller, David Lee. *The Poem's Two Bodies: The Poetics of the 1590 "Faerie Queene."* Princeton: Princeton UP, 1988.

Miller, Perry. *Errand into the Wilderness.* 1956. Reprint, Cambridge: Harvard-Belknap, 1984.

Miller, Shannon. *Invested with Meaning: The Raleigh Circle in the New World.* Philadelphia: U of Pennsylvania P, 1998.

Montaigne, Michel de. *The Essays: A Selection.* Trans. and ed. M. A. Screech. London: Penguin, 1993.

Montrose, Louis. "The Work of Gender in the Discourse of Discovery." In Greenblatt, *New World Encounters,* 177–217.

Mueller, William R., and Don Cameron Allen, eds. *That Soueraine Light: Essays in Honor of Edmund Spenser 1552–1952.* 1952. Reprint, New York: Russell & Russell, 1967.

Murrin, Michael. *The Allegorical Epic: Essays in Its Rise and Decline.* Chicago: U of Chicago P, 1980.

———. *The Veil of Allegory: Some Notes Toward a Theory of Allegorical Rhetoric in the English Renaissance.* Chicago: U of Chicago P, 1969.

Murtaugh, Daniel M. "The Garden and the Sea: The Topography of *The Faerie Queene.*" *English Literary History* 40 (1973): 325–38.

Nellish, B. "The Allegory of Guyon's Voyage: An Interpretation." *English Literary History* 30 (1963): 89–106.

Nicholas, Thomas. *The Pleasant historie of the Conquest of the VVeast India, now called newe Spayne, Atchieued by the vvorthy Prince Hernando Cortes marques of the valley of Huaxacac, most delectable to Reade: Translated out of the Span-*

ishe tongue, by T. N. Anno. 1578. Imprinted at London by Henry Bynneman. (Translation of Francisco Lopez de Gómara, *Istoria de las Indias.*)

———. *The strange and delectable History of the discouerie and Conquest of the Prouinces of Peru, in the South Sea. And of the notable things which there are found: and also of the bloudie ciuill vvarres vvhich there happened for gouernment. Written in foure bookes, by Augustine Sarate, Auditor for the Emperour his Maiestie in the same prouinces and firme land. And also of the ritche Mines of Potosi. Translated out of the Spanish tongue, by T. Nicholas. Imprinted at London by Richard Ihones, dwelling ouer against the Fawlcon, by Holburne bridge. 1581.* (Translation of Augustín de Zárate, *Historia del descubrimiento y conquista del Peru.*)

Nicholl, Charles. *The Creature in the Map: A Journey to El Dorado.* New York: Morrow, 1995.

Nohrnberg, James. *The Analogy of "The Faerie Queene."* Princeton: Princeton UP, 1976.

Orgel, Stephen. "Shakespeare and the Cannibals." In Garber, *Cannibals, Witches,* 40–66.

Parker, John. *Books to Build an Empire: A Bibliographical History of English Overseas Interests to 1620.* Amsterdam: N. Israel, 1965.

Parks, George Bruner. *Richard Hakluyt and the English Voyages.* American Geographical Soc. Special Publications 10. New York: American Geographical Soc., 1928.

Pearce, Roy Harvey. "Primitivistic Ideas in the *Faerie Queene.*" *Journal of English and Germanic Philology* 44 (1945): 139–51.

Penrose, Boies. *Travel and Discovery in the Renaissance, 1420–1620.* 1952. Reprint, New York: Atheneum, 1962.

Prescott, William H. *History of the Conquest of Mexico and History of the Conquest of Peru.* New York: Modern Library, n.d.

Quilligan, Maureen. *Milton's Spenser: The Politics of Reading.* Ithaca, NY: Cornell UP, 1983.

Quinn, David Beers. *The Elizabethans and the Irish.* Ithaca, NY: Cornell UP, 1966.

———. "New Geographical Horizons: Literature." In Chiappelli, *First Images of America,* 2:635–36.

———, ed. *The Voyages and Colonising Enterprises of Humphrey Gilbert.* 2 vols. Hakluyt Soc. 2d ser. 83–84. London: Hakluyt Soc., 1940.

Quint, David. *Epic and Empire: Politics and Generic Form from Virgil to Milton.* Princeton: Princeton UP, 1993.

Ralegh, Sir Walter. *The Discoverie of the Large, Rich, and Bewtiful Empire of Guiana.* Ed. V. T. Harlow. London: Argonaut, 1928.

Rathborne, Isabel E. "The Political Allegory of the Florimell–Marinell Story." *English Literary History* 12 (1945): 279–89.

Robbins, Keith. *Great Britain: Identities, Institutions and the Idea of Britishness.* London: Longman, 1998.

Rowse, A. L. *The Expansion of Elizabethan England.* New York: St. Martin's, 1955.

Said, Edward W. *Culture and Imperialism.* New York: Knopf, 1993.

Sherman, William H. *John Dee: The Politics of Reading and Writing in the English Renaissance.* Amherst: U of Massachusetts P, 1997.

Sidney, Sir Philip. *A Defence of Poetry*. Ed. Jan Van Dorsten. 1966. Reprint, Oxford: Oxford UP, 1993.

Sonn, Carl Robinson. "Sir Guyon and the Cave of Mammon." *Studies in English Literature* 1 (1961): 17–30.

Spenser, Edmund. *Book II of The Faery Queene*. Ed. G. W. Kitchin. 9th ed. Oxford: Clarendon, 1899.

———. *The Complete Works in Verse and Prose of Edmund Spenser.* Ed. Alexander B. Grosart. Vol. 1. N.p., 1884.

———. *The Faerie Queene*. Ed. A. C. Hamilton. London: Longman, 1977.

———. *The Faerie Queene: Book II*. Ed. P. C. Bayley. Oxford: Oxford UP, 1965.

———. *A View of the Present State of Ireland*. Ed. W. L. Renwick. Oxford: Oxford UP, 1970.

———. *A View of the State of Ireland*. Ed. Andrew Hadfield and Willy Maley. Oxford: Blackwell, 1997.

———. *The Works of Edmund Spenser: A Variorum Edition*. Ed. Edwin Greenlaw, Charles Grosvenor Osgood, and Frederick Morgan Padelford. Vol. 2. Baltimore: Johns Hopkins UP, 1933.

Steinberg, Clarence. "Atin, Pyrocles, Cymocles: On Irish Emblems in 'The Faerie Queene.'" *Neuephilologische Mitteilungen* 72 (1971): 749–61.

Swearingen, Roger. G. "Guyon's Faint." *Studies in Philology* 74 (1977): 165–85.

Tasso, Torquato. *Jerusalem Delivered*. Trans. Edward Fairfax. Intro. John Charles Nelson. New York: Capricorn-Putnam, n.d. (Also known as *Gerusalemme Liberata*.)

Taylor, E. G. R., ed. *The Original Writings & Correspondence of the Two Richard Hakluyts*. 2 vols. Hakluyt Soc. 2d ser. 76–77. London: Cambridge UP, 1935.

———. *Tudor Geography 1485–1583*. London: Methuen, 1931.

Teskey, Gordon. *Allegory and Violence*. Ithaca, NY: Cornell UP, 1996.

Thomas, Brook. *The New Historicism and Other Old-Fashioned Topics*. Princeton: Princeton UP, 1991.

Thomas, Hugh. *Conquest: Montezuma, Cortés, and the Fall of Old Mexico*. 1993. New York: Simon-Touchstone, 1995.

Tonkin, Humphrey. "Discussing Spenser's Cave of Mammon." *Studies in English Literature* 13 (1963): 1–13.

Treneer, Anne. *The Sea in English Literature from Beowulf to Donne*. London: UP of Liverpool, 1926.

Trexler, Richard. *Sex and Conquest: Gendered Violence, Political Order, and the European Conquest of the Americas*. Ithaca, NY: Cornell UP, 1995.

Whitney, Lois. "Spenser's Use of the Literature of Travel in the *Faerie Queene*." *Modern Philology* 19 (1921): 143–62.

Williams, Kathleen. "Spenser: Some Uses of the Sea and the Storm-Tossed Ship." *Research Opportunities in Renaissance Drama*. Report of the MLA Seminar 13–14 (1970–71): 135–42.

Williams, Norman Lloyd. *Sir Walter Ralegh*. London: Eyre and Spottiswoode, 1962.

Wittgenstein, Ludwig. *Tractatus Logico-Philosophicus*. Trans. D. F. Pears and B. F. McGuinness. Intro. Bertrand Russell. London: Routledge, 1974.

Woodhouse, A. S. P. "Nature and Grace in *The Faerie Queene*." A. C. Hamilton, *Essential Articles*, 58–83.

Yates, Frances A. *Astraea: The Imperial Theme in the Sixteenth Century*. 1975. Reprint, London: Ark-Routledge, 1985.

———. *Giordano Bruno and the Hermetic Tradition*. 1964. Reprint, Chicago: Midway–U of Chicago P, 1979

———. *Theatre of the World*. Chicago: U of Chicago P, 1969.

Index